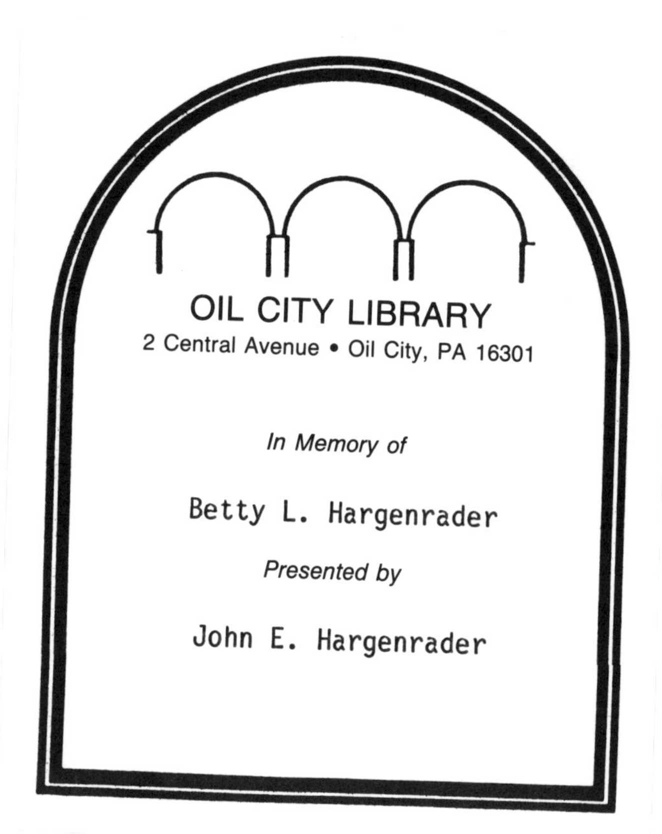

Fly Me to the Moon

Lost in Space with the Mercury Generation

Fly Me to the Moon :

Lost in Space with the Mercury Generation /

by
Bryan Ethier ;

Foreword by Wally Schirra, Mercury Astronaut .

Publisher's Cataloging-in-Publication
(Provided by Quality Books, Inc.)

Ethier, Bryan.
Fly me to the moon : lost in space with the Mercury Generation / Bryan Ethier —
1st ed.
p. cm.
LCCN: 98-68493
ISBN: 0-9653846-5-9

1. Popular culture--United States--History--20th century. 2. Astronautics and civilization. 3. Baby boom generation. 4. United States. Nationa Aeronautics and Space Administration--Public opinion. I. Title.

E169.02E85 1999 973.92
QBI99-151

Interior design and typesetting: Sue Knopf / Graffolio

Published by McGregor Publishing, Inc., Tampa, Florida

Printed in the United States of America

Dedication

To my wife, Debbie, for her patience, understanding,
and singular ability to listen to my incessant chatter about
rockets, capsules, and astronauts.

To my children:
Jordan, for reminding me what it was like to grow up
with space exploration. I see the stars in your eyes.

Brooke, whose silly expressions made me laugh
when deadlines made me cringe.
Like you, we should all laugh at life.

Adam, whose birth reminded me
how precious life is
and how remarkable a rocket launch can be.

Acknowledgments

To Lonnie Herman, for his insight. You allowed me to share with the world my vision of space flight. For that, I will forever be grateful.

To Dave Rosenbaum, my emotional compass on those days when I was lost.

To NASA's Laura Rochon, for providing me with the information and interviews necessary to make this book a reality.

Finally, I owe a debt of gratitude to: Dr. George Jutras, Christopher Glenn, Bob Crippen, Howard Benedict, Wally Schirra, Scott Pelley, Dr. Steve Eige, Father Tom Egan, Jay Egan, Jeannie Cunningham, David Crosby, Dayna Justiz Steele, Charlie Justiz, Dr. Lawrence Chapman, Gladys Lancon, Edith Balfe, Carol Heidhausen, Jack Kinzler, Sylvia Kinzler, Hank Hartsfield, Glendora Hill, Mark Lee, Steve Smith, Rick Chiavetta, Vern Estes, Mary Roberts, Matt Steele, Mark Bundick, Jay Apt, Gleda Estes, Mike Stanton, Jay Barbree, Senator John Glenn, Walter Cronkite, Gene Kranz, Alan Shepard, Gordon Cooper, Jim Oberg, Ellis Duitch, Don McKendry, Anne Marshall, and Bob Munkres.

This is your book.

Contents

Foreword

What a pleasure, when one can relive the past, and then advance to the future of the Space Age with the eyes and minds of the participants and the space buffs! This book does it all.

I am one of the early participants who, with some reluctance, became one of the Mercury astronauts. During my childhood, a generation before the author and his space buffs, I made model aircraft and visited Teterboro Airport in New Jersey to watch Clarence Chamberlain, Jimmy Doolittle, and others ply the sky. I went to the circus in New York City to see the animals, the trapeze acts, and some idiots shot out of a cannon into a large net. I did follow the adventures of "Buck Rogers" of the twenty-first century, but I was to be an aircraft driver, not a rocket man. After years of flying aircraft on and off aircraft carriers for the U.S. Navy, I entered Test Pilot school at Patuxent River, Maryland. It was there that I was taught to say "aircraft" rather than "airplane" (they do not plane).

After graduation, with Jim Lovell and Pete Conrad as classmates, we three and about sixty other test pilots were ordered to Washington, D.C. We were invited to take tests in order to be selected for the NASA space program. The finalists would be part of Project Mercury, and as astronauts, would climb into a capsule on top of a rocket to go into space! Where were those idiots who were launched out of that cannon at the circus?

Bryan Ethier, the author of *Fly Me to the Moon: Lost in Space with the Mercury Generation,* has captured the various thoughts and sensations of his generation, just as I had watched the Chamberlains and Doolittles of my generation. Bryan has done such a good job merging quotes from the spacemen with the feelings of his space buffs that I relived my experiences, but in addition learned what the space buffs of his generation were experiencing. He even delves into what his kids and their generation are thinking.

He brings in the reporters of the early space days: Cronkite, Bergman, Scott Pelley, Howard Benedict, and others, to tie the scenes together. I sat on the left hand of Walter Cronkite for all of the Apollo flights, from Apollo XI and on (no one sits on the right hand of Walter). Scott Pelley is really a member of Bryan's space buff club as well as a CBS type, as was Walter Cronkite. Walter is always available to tell how I felt on the launch pad for my Mercury flight. "Just think that these millions of different parts were assembled by the lowest bidder!"

You will discover many unknown people in this book, and they add so well to the time when we astronauts were unable to see or feel how other people were reacting to our adventures and to their own. It was an exciting time, and it is not over yet.

Wally Schirra
Captain, U.S. Navy (Ret.)
Former Astronaut

Prologue

I forged my way into outer space via a turkey baster, ball peen hammer, and a mangy, mud-brown, hooded winter coat that was tighter than a jet pilot's urine evacuation tube at 35,000 feet.

Clearly, these were not an astronaut's tools of the trade. I was no Wernher von Braun, however, the United States' German-born rocket scientist and the mastermind behind the Redstone rocket. I lacked the requisite scientific know-how. Still, my limitations as a design engineer didn't prevent me from constructing a spacecraft that would suit my needs, that would launch me hundreds of miles into outer space, away from my plodding day-to-day existence.

I never worried that my space capsule was neither constructed of titanium nor protected by a heat shield, as were the Project Mercury capsules. Liquid or solid rocket fuel? Didn't matter. Propulsion was never an issue: I was blessed with a firm right bicep and a windup that would have made Whitey Ford blush with jealousy. Did my rocket ship fly hundreds of miles into the star-speckled blackness of space? Nah. Better. My homemade Project Mercury capsule flew just as high and as far as my imagination could soar, and that was about as far as a Saturn V rocket could travel with a tail wind.

The year was 1962. Four years earlier, President Dwight D. Eisenhower had established the National Aeronautics and Space Administration, the government's watchdog over the nation's fledgling space program. 1958 was also the year I was born. Through its creation, NASA accorded a team of rookie astronauts the opportunity to shed the shackles of Earth-bound gravity. In turn, NASA also afforded me the chance to do what kids do best: imagine, dream, play with unfettered joy. To be whatever my heart told me to be, namely an astronaut.

My first journeys into space were brief and comparatively modest, not unlike those early suborbital flights of Mercury astronauts Alan Shepard and Gus Grissom. Space technology was still in its infancy, and I was just learning to write my name legibly. Since my annual space budget was a zillion dollars less than that of NASA, I had to suffice with the gadgets and odds and ends I was able to scrape up from around the house. Nevertheless, I was determined to build my own Mercury space capsule, even if it meant using such unlikely resources as Play-Doh and Lincoln Logs.

Fortunately, a bit of good luck rescued me from using such crude and ill-fitting components. One Sunday morning, as I watched my mother baste a turkey, it occurred to me that the eight-inch tapered baster she was using to bathe the meat in juice resembled a Mercury rocket. Moreover, the pliable, bell-shaped rubber bulb she used to squeeze the juice through the syringe looked just like a blunt Mercury capsule, minus its safety tower and escape rocket.

My mind began to yaw and roll with excitement, much the way von Braun's must have when he first envisioned a rocket powerful enough to take a man to the Moon. *Turn your back, mom,* I thought to myself. My mouth was dry with nervous anticipation, my hands moist with sweat. *Step aside and let me get to my future space capsule.*

By now, that common cooking instrument had miraculously transformed itself into *Freedom 7*, at least in my mind. Still, I had to get my trembling hands on it. From behind the wall that divided the kitchen and the living room, I waited for my opportunity to abscond with the rocket that would launch me into space.

And I waited . . . I waited longer than Alan Shepard and his aching bladder waited for relief inside *Freedom 7*. Finally, three Earth revolutions later, mom doffed her apron and headed out the side door. The baster lay unguarded upon the kitchen counter! I made my move. I raced past my toddling brother and sister, who looked at me quizzically, as if I were Captain Kangaroo in a space suit.

There! Finally, the missile was mine! But it was hot. The darned molten turkey juice which had solidified on the tip of the lousy baster was burning my hands! But I refused to let go. Holding the baster tightly to my chest, I stifled a cry of pain and raced off to the bathroom. Huffing and puffing with excitement, I slammed the door behind me and frantically separated the bulb from its gooey plastic nozzle. I secreted the nozzle deep in the garbage pail, then tossed the sticky, warm bulb into the sink. I had my *Freedom* 7, my very first space capsule; soon, I would embark on my maiden space voyage.

Like many youngsters born with the space program, I had wanted to be an astronaut since I was old enough to mumble the word *Sputnik*. I learned of our country's space heroes by sitting glued to our 16-inch mostly-black-and-sometimes-white Motorola every time a mission was in progress. I was barely old enough to recite the final 10 seconds of a countdown when Shepard became the first American hurled into space. Nonetheless, the images of Mercury space flights made an indelible imprint in my mind.

☆ ☆ ☆

During my childhood years, my favorite teachers weren't math, social studies, or science instructors; they were a revered reporter named Walter Cronkite and a dour-faced science reporter named Jules Bergman. To me, television and space flight went hand in hand, like John Glenn and ticker tape parades. Space travel was a visual thing: bell-shaped capsules and space men and gantries and smoke and fire and newsmen. Those fickle rockets rose dramatically, clutching at the sky as if caught simultaneously in the throes of life and death; then they plummeted slowly, softly into the blue-green ocean, to be scooped up by sea creatures called frogmen.

The astronauts' names whizzed past me like *Friendship 7* completing its second Earth orbit. I would never forget their names: Alan Shepard, Gus Grissom, John Glenn, Deke Slayton, Scott Carpenter, Wally Schirra, and Gordon Cooper.

During the Mercury era, space flight, for me, was about *space travelers* in *Way Out Machines*. It was about the stars and Mission Control. It was about weightlessness, and hearing one of those space travelers declare that the "clock has started; we are underway."

It was about all the things we could not physically experience—the heavens, God, the wonder of being *out there*—but a kid senses, in some unexplainable way.

Television brought the Mercury astronauts into our homes, our lives. But not into our souls. Gus, Alan, John, Wally, Scott, Deke, Gordon and I met in a place deep within my heart and imagination called the *Forever Little Boy*. It was there—in that part of us that remains forever young and wide-eyed innocent—that I and other junior astronauts were transported from our cozy, stable homes, to the tension-fraught launch sites at Florida's Cape Canaveral.

To fully experience Mercury, we had to *be* Mercury. So, from Shepard's flight on May 5, 1961, to Cooper's swan song on May 15, 1963, each four-year-old space fanatic in this nation scrambled like a demented Buck Rogers over and under couches, behind dad's workbench, into brother's '57 Mustang, in search of his own homemade Mercury spacecraft.

During those three historic years, I enjoyed the hygiene of a pampered aristocrat. I had reason to stay clean; our family bathtub served as my personal Cape Canaveral and Pacific Ocean in one.

It was a fitting place to practice liftoffs and splashdowns while enjoying complete privacy. You see, a space mission wasn't the sort of hobby you practiced before your family. Although space travel was new and exciting, parents (mine, at least) still considered soccer, baseball, and football more appropriate pastimes for their children. Doing splish-splash with a rubber kitchen instrument, well, that was weirder than thinking we'd one day fly to the Moon.

You had to be courageous to be a member of this Mercury Generation. Prejudice ran high in kindergarten, particularly among the benighted students. Not everyone was like I was, willing to endure epithets hurled at me from 41-pound bullies named Sally Ann. There was a price to be paid for spurning kickball to play flight director Chris Kraft within the sanctity of your own paint-chipped, puke-tan tub.

I did so, and with the same attention to detail as Project Mercury assistant flight director Gene Kranz. Not one of my missions was scrubbed, nor did any fail to achieve their goals. My *Freedom 7* invariably lifted off without a hitch. It even achieved low Earth orbit (defying the laws of science) and an apogee of about 37 inches, the height of my eyebrows.

Eventually, all these trips to the tub turned me into a wrinkled pink prune. I realized I needed another means to experience space flight. Thus the ratty-looking hooded coat.

The coolest thing about being an astronaut was that they let you wear those neat pressure suits. Any legitimate space traveler required a garment to protect him from the harshness of space. Moreover, it had to fit like a can around Spam. My un-spacelike coat suited my purposes, because it had a too-small hood which, when drawn tightly with frayed strings, fit snugly, like a space helmet. The hood was so tight, in fact, that it muted outside sounds. Now, I was really in my own universe. Only Mission Control could reach me through my *head-set*. Mom . . . dad . . . See ya!

Dad drove a black Ford Fairlane 500, one of a handful of spacious Fords he would own over the next 15 years. It had a huge back seat with plenty of floor space in which a space traveler could hide. Cruising in dad's Ford was like sticking John Glenn in a Space Shuttle and leaving him to fly it alone. I had my best space voyages on the nights we drove the 120 miles from our home in Guilford, Connecticut, to my birthplace, Woonsocket, Rhode Island. I spent the two hours huddled in the back seat—my *individually-molded Mercury couch*—engaged in two-way communications with my personal Capcom, or capsule communicator. The world outside was as dark as it must have been to our first astronauts. As the highway lights blinked past, and the stars twinkled, I felt I was 100 miles into outer space, floating silently, obliviously, around the Earth.

These unforgettable flights concluded when we reached my grand-father's house. Grandma and grandpa lived in a red brick ranch, com-plete with a cesspool that stunk to Mars. Inside the house a real *faux* ceramic fireplace, complete with real *faux* ceramic log, burned a real *faux* fire. If such a phantom conflagration did not soothe your soul, you could seek comfort from the many religious paintings, religious statues, and miles of rosary beads that gave their home a pious warmth. The place scared the hell out of me.

Despite the ubiquitousness of the Almighty, I found ways to escape when I was there. In the living room was a tattered gold recliner; it faced a broad picture window that provided a view of rows of houses from which no one person ever emerged. When it was dark, and the adults were in the kitchen screaming at each other over a poor bridge hand, I sneaked off to the living room and shut off the lights. I pulled myself onto the chair. Then, with a backwards shove, I was supine, like Shepard and gang, and staring through the window and into the heavens. This was a real Mercury capsule now, and what a thrilling vista it provided! Outside, the Moon was larger than life, all white and shimmering, and calling me. I tied my space helmet-hood, closed my blue-gray eyes, and drifted off into space. I remained in this delicious solitude until dad stuffed me back into the car for the long voyage home.

I lived this vicarious life throughout the Mercury, Gemini, and Apollo programs. Although by age 10 or 11, I still yearned to be an astronaut, I also dreamed of being the starting goalie for the Montreal Canadiens. And, when I was in fifth grade, did I ever want to kiss a girl named Lisa a thousand times!

By the time Project Gemini concluded in November, 1966, I had tossed aside the *Freedom 7* baster bulb. The ratty brown winter coat/space suit had been placed in storage. I was more mature now, more appreciative of the knowledge possessed by Cronkite and Bergman. The two giants of the space news world pulled me into NASA, inside the conference rooms, where orbits were discussed, where thrust was measured, fuel consumption was analyzed, and the electrical systems were examined in painstaking detail. As I watched my television space mentors I began to wonder, *How did the boosters work? How many parachutes did each capsule have? How did the engineers determine the exact reentry angle of the spacecraft?*

I perused bookstores and our public library and picked up every book on space flight that I could find. I collected space bubble gum cards, space pogs, model rockets, GI Joe Mercury Astronauts. And I watched every minute of every flight.

Nothing was more fascinating than a splashdown. Liftoffs? Von Braun and his predecessors had proved we could launch a rocket, but could we get one *back* from space? The possibility of a capsule being marooned in space haunted me like a phobia you can't explain and can't shed. It seemed so miraculous that this fragile Mercury vehicle, just 74.5 inches wide at the base and nine-and-a-half feet tall, could withstand the harsh void of space. Heck, to most kids, space was as enticing and foreboding as a 10-foot deep pool is to a new swimmer. Geez, to actually leave the safety of our planet, and make it back in one piece? Well, God had to figure into this equation somehow. To see the big, orange main chute billow out above the vulnerable pod kept me in a state of wonder for weeks after each mission.

How'd they do that? became my catch phrase.

Apollo was larger than life, because everything about it was big: the 363-foot tall Saturn V booster; the 'spacious' three-man command module. This was a mean rocket, capable of flying upwards of 25,000 miles per hour. Whereas Shepard had reached an altitude of about 116 miles, Apollo's astronauts would traverse some 240,000 interminable miles to reach the Moon.

Time: would it cease to exist on the Moon? I wondered.

In October, 1968, Wally Schirra, Donn Eisele, and Walt Cunningham flew farther, faster, and higher than any three men alive. Moreover, Apollo VII marked the first time live pictures were transmitted from a manned spacecraft. Five years earlier, I was forced to close my eyes to feel as if I were in space; now I could open my eyes and be in space, with real astronauts, who floated and bobbed and weaved in their weightless surroundings like ghosts in the still of night.

After watching astronauts wave to me from a world away, I had to have my own Apollo command module. By age 11, I realized that veterans of the Mercury Generation could no longer just reach into a toy box, pull out a cracked plastic widget, and call it *Charlie Brown* (the name given to the Apollo X command module). Certainly, I wished to emulate von Braun by constructing a capsule worthy of his name. But how? This wasn't a project whereby my father could assist; Mercury Generation junior astronauts always worked and flew solo. But to whom could I turn for inspiration and technical guidance? Walter or Jules? Textbooks? A mentor? God, the technology had become so complex!

☆ ☆ ☆

My ultimate undoing as junior rocketeer was precipitated by an underlying ignorance of my scientific limitations. I went to work building my Apollo command module facsimile armed with a relatively useless ball peen hammer, a handful of splintered two-by-fours, and a bag chock-full of broken knobs, dials, and switches. In retrospect, I should have examined my bill of materials. Right then and there I would have recognized my shortcomings. Nevertheless, my youthful dreams overwhelmed all sense of reason and logic. I went to work planning, designing, pondering. I put hammer to nail to board. Then I stopped.

Had I lost my head? I had no blueprints for this capsule mockup and no idea what would go where and why. How could I use straight, inflexible wooden boards to construct a vehicle whose cylindrical base tapered to a pointed nose? Where would I put the contour couch in which I, the astronaut, would lie?

Still, I forged on, slowly, stultified. For months, our basement was cluttered with dozens of components appropriated for my Apollo spacecraft. Each day, I padded down the 12 wooden steps only to find I was still an ill-prepared charlatan unable to proceed past the research and development stage of manufacturing.

If nothing else, my fruitless mission piqued my interest in the space race, all the way through Apollos VIII, IX, X, and XI.

From the comfortable confines of grandpa's tattered gold recliner, I watched spellbound as Apollo's star pioneers first orbited the Moon, then settled upon it. Apollo XI provided a sense of closure for us; we had fulfilled President John F. Kennedy's 1961 pledge to land a man on the Moon by the conclusion of the decade.

After the successful flight of Neil Armstrong, Mike Collins, and Buzz Aldrin, public interest in the space program began to wane, however. We had been there, seen that, done that, again and again.

A few weeks after the flight of Apollo XI, I returned to our basement. Feeling an unexplainable sense of loss, I packed up my ball peen hammer and tossed the broken knobs and cob-webbed boards into a corner with other bric-a-brac of broken dreams. My days as a junior astronaut were over.

☆ ☆ ☆

A generation has passed since I closed shop on my would-be rocket assembly plant. In 1978, I came within one one-hundredth of a GPA point of flunking out of college. My major? Urban Engineering.

I have long since stopped building ersatz space capsules, but my interest in space flight, and, in particular, Project Mercury, remains as keen as it was in 1963. Today I ask myself: Where would we be had there not been a Project Mercury?

It amazes me that 35 years have passed since Gordon Cooper's *Faith 7* flight marked the conclusion of Project Mercury. We lost Gus Grissom, along with Roger Chaffee and Ed White, in the 1967 Apollo I fire. In 1993, Deke Slayton succumbed to cancer. In 1998, Al Shepard passed away after battling leukemia. The four surviving astronauts are now in their 70s, yet they remain active in society and interested in space flight. Thirty six years after circling the Earth three times, Glenn returned to flight, this time aboard a roomy Space Shuttle.

Sadly, time does prove to be linear when you return from space. As Glenn once told me, "With all the research we've done, we still haven't found a cure for the common birthday."

Books, tapes, and photos keep them forever young to us, and their exploits forever real. Those of us in the Mercury Generation know, however, that when we need to really connect with the Original Seven, all we have to do is look inside our hearts, and we will return to a time when we all were wide-eyed boys and girls wondering about worlds beyond our own.

Heroes

She learned about the stars from a popular animated movie about the king of all lions, blessed with the roar of a thousand thunder claps and the wisdom of Merlin. Years earlier, I had ventured into the world to experience the hypnotic beauty of those same beacons. Staring into a cold, black sky, filled with winking stars, I wondered, *Who's out there? Man? Martian? God?* I vowed to one day discover firsthand the secrets of the heavens.

In her carefree world of imagination, enlivened through the magic of VCRs and video tapes, she learned that the spirits of the mighty lion kings watched her from those stars. They were her guardian angels.

I never tried to disabuse her of that notion. She was just two weeks shy of her fifth birthday, all blue-eyed innocent, eternally naive and impressionable. Her silly notions of the stars were no more ingenuous or farfetched than the ones many people entertained before we took our first shaky steps off our planet.

When I was her age, however, I had already begun to learn the lessons of the stars. Over the next 35 years, I concluded that these celestial bodies were extensions of my soul. When I gazed upon them on cloudless nights, they whispered *You can do anything you imagine.*

Now, nearly 39 years old, I had resigned myself to the fact that I would always be a dreamer, a person who saw the stars and life as he wanted to see them. Perhaps that's different from seeing the world the way it really is.

I mused over the stars and my precocious daughter, Jordan, on a most breathtaking evening in June, 1997. The weather was mild, free of the oppressive humidity that typically weighs down a New England summer. The Ethier family—that is, my wife Debbie, and daughters Brooke and Jordan—had eaten outdoors, on the deck of our split level ranch, huddled around a too-small plastic outdoor dining set. I studied Jordan as she stared skyward.

"Jordan, what do you see above you?" I asked, chilled that she might be thinking about the stars, as I often did. Beside her, Brooke, her two-year-old sister, picked at a knot of starchy spaghetti, then cast a look of curiosity at her big sister.

"An umbrella," Jordan replied.

"No, what do you see above the umbrella?"

"The trees," she responded, her impish face bright with satisfaction.

Debbie reached across the table and tapped Jordan on the hand. "Above the trees, Jor," she corrected.

As if on cue, Brooke and Jordan peered further heavenward. I searched their faces; their eyes were dim, distant. *Do they see something only children can perceive?* I wondered.

Before they could answer, the bark of a neighbor's dog diverted my attention to the southeast corner of our yard. The horizon was splashed sapphire blue, with the water of the Long Island Sound. The setting sun had cast deep, dark shadows across the perimeter of our yard; the jungle of trees and vines was so green, *Technicolor green,* the yard looked as if it were a production set from *The Wizard of Oz.* This giant wall of olive was dotted with speckles of dove white from a dogwood tree that seemed juxtaposed in this tropical setting.

I shook my head to clear the mental cobwebs.

"Do you see the Moon, Jordan?"

"I saw one this morning when we were going to school." Her head vacillated as she searched the sky.

"That's right. Did you know that some men actually flew to the Moon in a big rocket?"

Her pool blue eyes narrowed. A studious look passed over her.

"Huh?" she murmured.

Debbie glared back at me. Her eyes searched mine, and asked, *What are you trying to do*?

Suddenly, the tension of the moment was broken when an ant the size of Gemini III bounced onto my arm.

"Daddy, look, an ant jumped on you," Jordan cackled, forgetting about the sky.

"Ant, ant!" Brooke chimed in.

Annoyed at this interruption, I flicked the ant aside with a wave of my hand. My stomach was beginning to tighten with frustration. *Be patient*, I told myself. *Jordan's only four years old.*

I continued. "Jor, did you know that men actually walked on the Moon, the way we walk on the ground here?"

Jordan's face widened into an expression of doubt.

"Why did the ant crawl on you, Daddy?" she responded, still obsessed with the bug.

"She's only four," Debbie chided, sensing my frustration.

"But when I was four years old, I was watching men fly in space!" I barked. "I had already fashioned my first space capsule out of my mother's turkey baster. Whenever a mission was on, I was in front of the TV set watching every minute of that mission. This generation . . . It doesn't even know what a space capsule is. If, and I mean if, they see a Space Shuttle lift off, it's on a 23-second news clip on a cable news station. Boy, have the generations changed."

"Space flight has become so common these days, no one gets excited," said Debbie, echoing the sentiments of countless others.

I rolled my eyes. "Did you realize that just a few days ago we landed a robot on Mars? Today, we scan the newspaper headlines and see pictures so crisp, so full of this planet, it's like we've just taken a snapshot of the Mojave Desert."

"I understand that," she said.

I turned back to Jordan. "Jor, do you know about the planet Mars?"

Her face lit up.

"Yeah," she cried. "I learned about that on the *Magic School Bus*. From Arnold."

"That's right!" I shouted triumphantly. The *Magic School Bus* was one of those typically clever animated public television shows that taught children how to add, spell, take their temperature, fly to the Moon. . . .

Cheerfully, Jordan helped her sister from the dinner table. Together they raced off into the living room.

"We're going to watch a movie, Dad," she reported.

A *movie*. . . . Perhaps they'd learn a second language, or learn how to bone trout table-side, like a skilled waitperson. Where would Jordan and her friends be without VCRs and Walt Disney? Times had surely changed in the many years since I first mumbled the word *astronaut*.

Debbie stood up from the table and waved her hands back and forth before my face. "Hello!" she said sarcastically.

Huh? Yeah. Jordan. Family. Dinner. Reality.

"You can't expect a four-year-old to understand all this stuff about space," she said. "I've learned more about it in six months than I ever knew before."

But I did expect my four-year-old daughter to be curious, to step outside into the hush of night, to ponder those specks of light amidst all the darkness. I wanted her to challenge herself, to create her own questions, to find her own answers. TV all too often served as her teacher.

As a child, my life was not nearly as canned as are the lives of today's generation. But my circumstances were different, back in the early 1960s: I belonged to the Mercury Generation. I peered into the world and saw what I *wanted* to see, not what was actually there, or what was being shoved down my throat by an electronic medium. To me, the Moon did more than hang in the sky like an ornament; it beheld the secrets of space exploration, and a road map to the stars.

I had been born into the Mercury space program. I grew up with the program and with space flight. To a four-year-old, life is big and impressive enough; for me to see the first Americans fly in space, at such an impressionable age, would forever make NASA, astronauts, and space voyages appear larger than life. Even as an adult, I would see Moon Men through the eyes of that four-year-old little boy.

At age four, I knew I wanted to be an astronaut. I watched every minute that I could of the first Mercury flights, even though kids my age had attention spans that lasted as long as a TV commercial.

Jordan wanted to be a firefighter. Noble work. Not the type of career, however, that enlivened the dreamer in a person. At least that's what I thought. Dreaming... a requisite for all members of the Mercury Generation. Mercury was about heroes, too, as were all the manned space programs. Jordan saw a burly man wearing a long black coat, big black and yellow boots, and yellow hard hat rescuing a cat from a burning house and considered him a hero. She was right, of course. But when I considered heroes, I thought of pioneers such as John Glenn, Alan Shepard, Gus Grissom, Deke Slayton, Wally Schirra, Scott Carpenter, and Gordon Cooper. These were ordinary men doing truly extraordinary things.

These were individuals brave enough to face the unknown, an act of courage often beyond my reach.

After a while, Debbie went inside the house to wash the dishes. I remained outside and reflected upon the accomplishments of the seven men who helped shape my life.

They were in their 70s now. It pained me to think that despite their superhuman accomplishments in outer space, these were mortal men. Soon they would all be gone, for outer space was no sanctuary from death. You could try to out race it at 25,000 miles per hour, you could try to hide from death on the Moon. It didn't matter, though.

What happens when they are all gone? Will a part of my soul go with them? I wondered.

It seemed so hard to believe that by the time the millennium dawned, we would likely lose another one of my icons. We lost Gus in the 1967 Apollo I fire that also claimed the lives of astronauts Roger Chaffee and Ed White. Pilots got killed flying their craft; that was part of the gamble they took every time they climbed into the cockpit of an X-15 or an Apollo command module. But when Deke succumbed to cancer in 1993, that touched all of us on a fundamental level. You almost expect an astronaut to go down with his capsule, like a captain with his sinking ship. But *cancer?* Astronauts, I began to divine, were as transient as space flights.

The traits that had made the Mercury 7 astronauts heroes for me as a kid made them role models for me as an adult. Their exploits proved to me that with perseverance, effort, study and faith, I could accomplish anything. But I was lucky, I was a member of the Mercury Generation. I grew up in an era when space flight was a metaphor for life: *If we could leave the planet, we could do anything.* Glenn and his colleagues helped mold our attitudes and beliefs by proving that self-limitation did not exist, if you just believed.

☆ ☆ ☆

On a dreary afternoon in April, 1997, I received a call from Senator John Glenn.

"Hi, Bryan, it's John Glenn," he said, amicably. "I want to apologize for having to postpone our interview."

Isn't that neat, I thought. Here he is, a U.S. Senator, kind enough to consider the feelings of a writer with whom he has never spoken. But John Glenn has always been this way: thoughtful, and cognizant of his public standing. In fact, most of the astronauts—retired and active—are just as accessible and amiable.

Glenn says that's because the astronauts recognize the impact they have had on people. "I have traveled all over the world and it has even surprised me how many places I go where people remember those early days and the part we had in them," he said. "It's heartwarming in some respects, and it's amazing in others, to think that what we triggered off back there with our efforts started something the whole world has been interested in."

For this Democratic senator from Ohio, every day is the flight of *Friendship 7*. Hardly a day passes when a 13-year-old kid with black-framed glasses, red crew cut, and freckles smearing his bright cocoa face doesn't ask what it's like to be an astronaut.

That is the moment when Glenn's anecdotes launch his listener into space. It is there that time becomes suspended and Glenn and audience become pals.

Glenn takes the time to chat with his fans because it promotes space flight and fosters interest in a program that has become an afterthought to many.

"I feel there's a responsibility to still participate in it, if we can still do some good," he said. "If the kids come to the office here [Capitol

Hill], or if I run into them on the subway and they want to stop and talk a minute, I don't hesitate to stop and talk. I think it's good. I think that's almost a duty we have."

One could argue that Glenn and his colleagues more than fulfilled their obligation to the public. Not only did the astronauts successfully accomplish the goals of Project Mercury—to launch a man into space and return him safely to Earth—they also revived the flagging spirit of a country whose self-esteem had been rocked by the technological advancements of the nation's rival, the Soviet Union.

Because of these first small steps into space, we now have the Hubbell Space Telescope, a marvel of technology that allows us to gaze around the universe. Then there's Pathfinder, which transmitted to Earth pictures of a Mars landscape so vivid, it's as if we are there, creeping and crawling around those dun-colored rocks.

Then, of course, there are the many satellites in space that bring us telephone service, television, and countless other services.

All this and more, thanks in part to Project Mercury.

☆ ☆ ☆

America's first rides into outer space began with a communion, April 9, 1959, in the auditorium of a Washington, D.C. hotel. There the world met the seven fearless men who would take us to new worlds. They looked like anything but space voyagers, however; they were neatly-groomed, dressed smartly in suits as stiff as the expressions on their faces. All were in their 30s, and all boasted impressive records as military aircraft test pilots.

Not one, we concluded, had his wits about him. They were going to what? Sit atop an explosive rocket that would launch them into space? Why not just stick a lit firecracker under their butts and send them to Kingdom Come? Sure, from the pulp science fiction magazines, astonished people had read about the possibility of space exploration. But to think of actually sending a human being into an environment we had never explored was tantamount to suicide!

Fortunately, the seven men chartered to do the job thought otherwise.

Getting into space was manageable, so NASA claimed; the astronauts had faith in the technology and in their skills. Surviving this grueling press conference was another matter.

The staccato click-click of cameras filled the buzzing room. Teams of reporters, most clad in white shirts and dark ties, weaved in and out like moles to get the perfect snapshot of these heroic explorers. *Who are these men?* the press wondered.

A name card placed on the table before each astronaut disclosed his identity: Navy Lieutenant Malcolm Scott Carpenter; Air Force Captain Leroy Gordon Cooper; Marine Lieutenant John Glenn, Jr.; Air Force Captain Virgil I. "Gus" Grissom; Navy Lieutenant Commander Walter M. Schirra, Jr.; Navy Lieutenant Commander Alan B. Shepard, Jr., Air Force Captain Donald K. "Deke" Slayton.

They would be our Magellans.

These test pilots-turned-star-voyagers accepted this unprecedented challenge because it offered them the supreme reward: they would fly higher, faster, farther than any men before. In exchange for this honor, the seven astronauts would be our nation's ambassadors in outer space.

On October 4,1957, the Soviet Union had shocked America by launching the world's first artificial satellite, *Sputnik I*. The steel sphere was only 23 inches in diameter and it weighed just 183 pounds. It could have been a 10-ton nuclear bomb for all we cared, however. The droning *beep-beep-beep* radio signal *Sputnik* transmitted back to Earth could be captured by ham radio operators from New York to Los Angeles. For the typical American citizen, *Sputnik* underscored our greatest fear: The Soviets, our bitter enemies, now controlled the heavens, and possibly our lives. To President Dwight D. Eisenhower, *Sputnik* proved the Soviets' superiority in space and rocket technology. Scientists had determined that it would have taken a rocket with 400,000 pounds of thrust to launch *Sputnik* into space. The United States' most powerful rocket, the Atlas, could amass only 360,000 pounds of thrust, and it was not yet operational.

The success of *Sputnik* was also a shot to the nation's ego.

"Americans had considered themselves preeminent in science and technology, and we were," recalled Glenn. " All at once they put this thing in orbit when we were still trying to get ready. This shocked us, because it was the days of the Cold War and we felt the Soviets were doing their level best to sell communism around the world. And here they were touting—they were flaunting—their preeminence in space as evidence that their system of government was better than ours."

The ramifications of the launch of this tiny satellite were staggering. Eisenhower feared that if the Soviets had rockets powerful enough to launch a satellite into orbit, they might also have the lifting capabilities to drop a nuclear warhead onto our beaches, our military bases, our backyards.

Red-faced, Eisenhower went to work, accelerating the nation's research and development of long range missiles.

To Gene Kranz, assistant flight director in Mission Control and section chief for flight control operations during Project Mercury, the early U.S./Soviet space rivalry was a "battle for the hearts and minds of the free world."

"This was real warfare we were involved in," he said. "It wasn't the 'Red Scare,' it was more fundamental than that: We were second best and Americans don't like that kind of role, that kind of position."

Yet, Americans would have to settle for a silver medal in space technology, at least for a while. In November, the Soviet Union launched *Sputnik II*. On board was the world's first traveler, a dog named Laika.

In 1958, Eisenhower established the National Aeronautics and Space Administration, a civilian agency charged to govern the nation's space policy. NASA's first project would be to launch a man into space and return him safely to Earth. In October, Project Mercury was approved to carry out that mission.

But who would fly these mighty rockets into a fate less certain than life itself? In the fall of 1958, NASA began setting the criteria by which the initial group of astronauts would be selected: The first star travelers would be men; they would have to be courageous and strong to bear the mental and physical rigors of flight.

Ultimately, military test pilots were chosen. They could be no more than 40 years of age, no taller than 5'11", no heavier than 180 pounds (to enable them to fit into the tiny capsule). They required a formal engineering degree, or its equivalent. Eisenhower and NASA also reasoned that a pilot's thousands of hours of flight experience would serve him well when he encountered the many unknowns of space flight.

From an original list of 508 test pilots, 32 volunteers underwent weeks of medical and psychological tests; they had their minds probed, every inch of their bodies inspected. What wasn't nauseating was annoying, particularly the tests and interviews administered by expressionless psychologists.

Finally, some three months after the outset of the process, the seven men who would carry the nation's banner into space stood before the world and a sea of reporters. And a nation completely swept away with anticipation.

"There was a sense of excitement in the Mercury days," recalled Shepard, prior to his death in 1998. "That has not been repeated for two reasons, I think. One was the sense of something really spectacular and brand new with the excitement of leaving the planet for the first time. The other was the sense of national pride, because it was obvious we were competing with the Soviets."

From the very first clicks of the photographers' cameras that historic April day in 1959, the Mercury 7 astronauts were deemed celebrities and heroes.

"We were at first extremely surprised when we were announced to the whole world, and how crazy everybody went over the whole thing," Cooper said, laughing. "We just couldn't believe the whole country would go so gung-ho over the situation."

Schirra captured the essence of the nation's excitement: "Any time you leap on top of a rocket and go into space, that's a hell of a big deal."

It *was* a big deal, for us, for much of the media. For the people who had not logged countless hours in jet aircraft. Getting these seven pioneers to share our wide-eyed exuberance was another matter. These were stick and rudder guys, accustomed to stepping into the cockpit of a potentially deadly aircraft. A Mercury flight would be similar. The real challenge was bridging the gap in perspective between astronaut and ogling Joe Public.

Glenn helped ease this anxiety during the brief question and answer period of the press conference. When asked by a reporter which of the medical tests was the worst, the smiling, golden-haired and freckle-faced Glenn responded sheepishly: "It's rather difficult to pick one. If you figure out how many openings there are in the human body and how far you can go in any one of them, you answer which would be the toughest for you."

Thanks to Glenn's self-effacing wit, the room broke into laughter and the self-conscious astronauts warmed to the cameras and the nation.

"I think the thing that made them magical was they built them up as going through this incredible selection process," said Mark Lee, today one of NASA's crop of Space Shuttle astronauts. "I think people had this image of them being the most incredible pilots, engineers, athletes we ever had."

When the curtain closed on that introductory press conference, the Mercury 7 astronauts had left their mark on the nation.

They had given us a collective goal more grand than any we individually embraced. More important, they had given each of us a sense of hope: The prospects of these men leaving our world made the long days in the oppressive textile mills bearable for granddad. It gave the unemployed reason to believe that if we could fly in space, maybe one day everyone would have a good job.

Mercury represented our boldest dreams, pitted against pragmatic, relentless reality.

"They inspired a whole generation to want to be a part of the space program whether it was an engineer or a secretary, they just wanted to be a part of it," recalled Howard Benedict, who covered the space program for the Associated Press for more than 30 years.

Over the next four years, space flight would miraculously transform us from mother, teacher, salesman, construction worker, into John Glenn, Gus Grissom, Alan Shepard, Wally Schirra, Scott Carpenter, Deke Slayton, and Gordon Cooper. From liftoff to splashdown, company time clocks magically stopped ticking.

"I think the space program has given people, since Mercury, a sense of 'no, this isn't all there is'," said Scott Pelley, the CBS News Correspondent who has covered more than 20 Space Shuttle missions. "In a real world, here and now way, there are worlds to explore, other places to be, new things to learn, and we don't know everything there is to know. It gives people a fresh perspective on their own world, knowing there is much more to life than seeing what is right in front of us."

Everything had changed, even the trivialities of day-to-day existence were more intriguing. Take walking the dog, for instance. Now you and Spot could venture into the still, spring night and wonder if we would ever travel to the distant stars.

Re-reading the old science fiction classics elicited a chill: Life had become Science-Fact.

But there was one problem: We actually had to do this! We had to take a man, shove him in a can they called a capsule, stick astronaut and capsule on top of a machine filled with explosive liquid propellant, and launch them out of the atmosphere. Poke a hole through an envelope called gravity. Way out of sight.

Of course, we believed that NASA and the subcontractors charged to build the rockets and components knew precisely what they were doing. We forgot, however, that no one had a blueprint for flying to the stars. Who knew about Zero-Gravity, G-Forces, Reentry, Liftoff, and Splashdown? Then there were the notorious rockets, the Atlas and the Redstone. The Atlas, the only U.S. booster powerful enough to launch a payload into Earth orbit, had displayed an annoying propensity for going *KABOOM*.

"It was really a step by step build-up to get to the Moon," said Cooper. "There were so many things we didn't know about. Everything was a learning process, and 'cut and try to see if it would work' procedure."

Atop these missiles of peril, a human being would literally fold himself into the 9½ foot by 6 foot capsule, his home in space. The astronauts would refer to the capsule as a "garbage can," for it offered a similar paucity of aesthetics and even less leg room.

"Mercury was a challenge," said Kranz. "But no matter how inept we were, and we were inept, I think the people who actually designed, tested, and flew those missions recognized how far we had to go to be reasonably competent. To put it bluntly, we didn't know what we were doing in many areas of the Mercury Program and we were fortunate our country understood no achievement came without risk and with no guarantees."

To that end, for the next two years, the astronauts familiarized themselves with the miles of wire, hundreds of meters, relays, switches, and connectors that composed the capsule and rocket. They monitored the development of the spacecraft from heat shield to booster to escape tower, often visiting the plants of the subcontractors which manufactured the components and the sub-assemblies. The crew suggested changes in design, such as the introduction of a capsule window and removable hatch. When not poring over volumes of technical manuals, the astronauts subjected themselves to further rigorous training, often in devices and machines that resembled the amusement park rides renowned for making a nine-year-old lose his lunch. They did all this to provide answers to questions about space flight that were yet to be posed. For 24 hours a day, they did little but train to be astronauts.

In time, the astronauts, subcontractors, and NASA engineers became a team.

"We learned to check our egos in at the door," said Kranz.

Notwithstanding the cohesiveness of this expansive group, the United States still lagged behind the Soviets in the space race.

By the end of the 1950s, the Soviets had not only launched a dog into space aboard *Sputnik II*, their space probes had even photographed the far side of the Moon. America, conversely, had launched its first satellite aboard Explorer I on January 31, 1958. It had also sent a space probe, Pioneer 4, to within 37,300 miles of the Moon, and had launched the first primates into space, monkeys Abel and Baker, on May 28, 1959.

The Soviets upped the stakes on August 19, 1960 by safely recovering the first orbited animals from space. The U.S. again countered by launching the chimp Ham into space on January 31, 1961.

While the flights of animals proved to be a prudent step in the gradual process of sending a man into space, it nevertheless did little to excite the nation. Americans wanted Buck Rogers, not a diaper-clad banana-eating monkey who responded robotically to commands like Pavlov's dog.

They got their wish . . . almost.

A space voyager did step forward, but an astronaut he was not. On April 12, 1961, Russian *cosmonaut* Yuri Gagarin catapulted the Soviets a galaxy ahead of their struggling opponent by doing more than merely riding into space. The first man to rocket toward the heavens also orbited the Earth once during his 108-minute flight aboard Vostok 1.

The nation was stunned, all over again.

"We had tried and hadn't made it before," said Glenn. "We had been on and off and on and off and then all at once they did it. And they didn't do it with a suborbital, they went straight to an orbital flight."

The Soviets performed this Herculean feat under a cloak of secrecy. Had Gagarin's flight failed, or worse, had he been killed, no one would have been the wiser.

Had his flight failed . . .

While the Soviets were gloating over their major technological victory, the United States continued to suffer political and public relations disasters. On April 17, 1961, the aborted Bay of Pigs invasion shot a hole right through the country's soul and tarnished the image of its new, young president, John F. Kennedy.

On that date, some 1,300 Cuban refugees armed with U.S. weapons landed at the Bahia de Cochinos (Bay of Pigs) on the south coast of Cuba. They intended to cross the island to Havana, where their mission was to overthrow the government of Fidel Castro. Without the support of the local population, however, they were quickly turned back by the Cuban army. Ninety were killed, the rest captured and taken prisoner.

Kennedy and the U.S., already embarrassed by their inability to overtake the Soviet Union in the space race, were further humiliated by this debacle. It was time for a hero to step forward and set sail.

But who would be the pilot chosen to ride atop the ticking bomb? The seven astronauts had all excelled in their classroom studies, training, and engineering. All were bright, articulate, athletic, eager to be the first American in space. Most important, despite sharing a competitive fire, they had worked well together as a unit, with a single purpose in mind.

NASA chose 37-year-old Alan Shepard to make the historic flight. Unlike Gagarin's mission, Shepard's would be covered live on television and radio, for the entire world to witness. Shepard would not be the first man in space; but he would be the first free man in space, on May 5,1961.

Hundreds of hours of training in capsule simulators had blunted much of Shepard's anxiety about his impending flight. But members of the Mercury Generation fretted as if *they* had entrusted *their* fate to a rocket that as yet had only hurled animals and satellites into space. As the day of Shepard's flight drew agonizingly closer, we found ourselves thinking a little less about the office, PTA meetings, grocery shopping, even our plastic blow up punch clowns.

Shepard's mission became our own.

☆ ☆ ☆

On the night before his flight, Alan Shepard slept soundly. The same could not be said for many of us, however. Fourteen-year-old Jim Oberg was among those fretting over the fate of the astronaut and the rocket. It was like Christmas Eve, although Oberg's dreams were not filled with images of presents peeking out from underneath a tree taller than the sky itself. His mind's eye saw violent thunderstorms scrubbing the mission; it saw the mighty Redstone rocket rise a couple of feet off the pad, then topple over in a fiery earthshaking explosion. Jim Oberg slept restlessly and envisioned the worst.

He had grown up reading about space flight like others of his generation. Like the many kids enamored of rockets, Oberg tried his hand at building his own homemade capsule from a refrigerator box. He fashioned knickknacks into aviation instruments such as tachometers, barometers, and altimeters. Then he and two or three co-pilots teamed up for their own journey into space.

Sputnik had satisfied a part of him that yearned for science fact to catch up to science fiction. The launch of *Sputnik I* had also filled Oberg with a tremendous sense of fulfillment: The once fancied notions of space exploration were now reality. Best of all, no one laughed at him when he brimmed with excitement over the space program.

Most of us were doing the same.

Oberg and millions of Americans were still tossing and turning and talking in their sleep when Alan Shepard arrived at Launch Complex 5 at 5:15 a.m. Clad in a thick, tightly fitting silver pressure suit and white helmet, Shepard looked every bit the extraterrestrial as he emerged from a NASA transfer van. In front of Shepard, the launch site was awash in milky white light, as was the rocket, which was a giant, shimmering white and black pencil, issuing pillows of coolant from its core.

Shepard panned the launch area, then he took a few steps toward the Redstone. He gazed upward at the rocket, which stood some 83 feet tall. *Well, there's an old friend,* he thought. *I'm not going to see it again; let's take a quick look.*

Then he was gone, rising inside the gantry elevator that would take him up to *Freedom 7.*

As Shepard and Mercury Control's flight team reviewed the mission flight plan, the rest of the nation awoke to a day unlike any other. *Hoot, today we're going into space!* We raced to our dinky black and white TVs, or to our radios, to get word on the flight. Breakfast was toast jammed nervously into mouths dry with excitement.

Boys and girls from Alaska to Maine ruefully raced to catch their morning school buses, which threatened to leave without them. Shouts of protest trailed in the distance: "Ma, must we go to school today?" Dads shaved carelessly—*Damn, cut my chin*—and nearly choked themselves to death with Windsor knots as tight as hangman's nooses. "Hey, boss, do we have to come to work today?"

Moms everywhere, clutching cloth diaper in left hand, child in right, flipped anxiously between TV channels to get the best coverage. Meanwhile, their babies cringed at the sight of the dreaded diaper pin weaving around their fannies like heat-seeking missiles approaching the tail of a jet aircraft. *Not my butt, Mommy, please. Stick the pin in Susie's fanny. She didn't eat supper last night.*

High school kids skipped off to school, transistor radios in hand. For once, rock and roll was not blaring back at them.

In Woonsocket, Rhode Island, 16-year-old Jay Egan stared at a snowy image of *Freedom 7* on the screen of his parents' television set.

Egan would have climbed right through the television and folded himself into the capsule, beside Shepard, if it were feasible. He had for years been enthralled by the prospects of space exploration; growing up, he had read every book, every article he could find on NASA, the rockets, the astronauts. He had devoured the television documentaries on Shepard and crew, his overloaded senses scarcely getting enough of the fascinating information.

Even on the flights when monkeys had whirled around the Earth, Jay Egan had waited impatiently for the evening news to chronicle all the details of the flight. Then, suddenly, he was a statue before his TV. *One day I will be an air force test pilot,* Egan had vowed.

But now, in the back of Jay Egan's mind, he knew he never would. His eyesight was poor, and he knew in his heart that no branch of the service recruited pilots who could not distinguish enemy from allied aircraft.

Oh, well. Such is fate, he sighed. But even if he lacked the wherewithal to be a test pilot, he still could *fly.* Up with Shepard, up with the other six astronauts. That's why Jay Egan was today a member of the NASA team; this was the winning team, the team that gave the world hope for a brighter tomorrow.

The agency can do no wrong, they are the golden boys, thought Egan. *So are the astronauts. God, they are putting their careers, their lives, on the line for the nation, for the future of the space program. And it hardly seems to affect them.*

To Egan, that's what made Shepard and his flight mates heroes: They set a goal and pursued it with unflagging confidence and fervor.

The race to conquer space had grown into something larger than just a jealous competition between titan nations, Egan concluded. Today, the day of Shepard's flight, was the Super Bowl. We had carried the ball and had fumbled so often, we were due to score a touchdown. So would Shepard's flight be that grand score? Egan hoped so, for the country had been in a blue funk since Gagarin's flight.

We're the home team. We'll come through, I know it, he thought.

Egan's thoughts raced throughout the galaxy, as did the contemplations of anyone even remotely interested in Shepard's flight. Would America watch as its hero was blown to bits on the launching pad?

We all closed our eyes and gritted our teeth. Egan forced himself to think positively. He sat on the floor beside his 18-year-old brother, Tom. Mom and dad were riveted in comfortable chairs, and Aunt Fu Fu milled about the spacious living room of this 150-year-old house like a skittish cat fearing danger. Aunt Fu Fu flipped between channels, Jay and Tom Egan flipped between channels. Everyone wanted to get every bit of minutiae they could about Shepard's condition.

So the Egans sat and waited, afraid to leave the room for fear of missing some vital piece of flight information. We were all a part of Shepard's support team now, and we did not dare break the spiritual link between ourselves and the astronaut. Bathroom? Screw it, we'd pee in our pants, as Shepard would during his lengthy wait in the capsule.

Time ticked by in half-seconds, rendering us emotionally exhausted. Finally, when we could no longer stand the waiting, there were just seconds left. FIVE, FOUR, THREE, TWO, ONE . . .

No turning back. With our eyes closed, our fingers and toes crossed, we felt ourselves begin to shake inside as the Redstone trembled and then roared to life. At 9:34 A.M. EST, *Freedom 7* leaped off the pad with a spurt, like a model rocket in your backyard, hardly affected by the clutches of gravity.

Thousands of infants nationwide screamed in protest as moms unintentionally punctured their pink fannies with harpoon-like pins. Not one high school student cared a lick whether an isosceles triangle had two or three equal sides; each was too busy watching a man ride a rocket into space.

We grabbed a friend's hand for emotional support as Shepard reported back to Mercury Control: "Liftoff. The clock has started. Yes, sir, reading you loud and clear. All systems are go!"

Back at the blockhouse, Capcom Gordon Cooper peered around the busy, emotion-filled room. Not a dry eye in sight. Even Wernher von Braun and his team of stoic German rocket scientists had been reduced to misty-eyed little boys, so proud of their creation, the Redstone booster.

"Go!" Jay Egan shouted. "Go!" the rest of the nation cried in unison.

Suddenly, Egan's world was rolling and yawing, tossed upside down, backwards and around like a capsule out of control.

Reality had gone haywire and it was almost too much for him to bear.

Other space maniacs raced around their living rooms, leaping in celebration, hugging, kissing. Those driving by beeped their car horns and pulled over to the side of the road, too overwhelmed to drive. The youngest members of the Mercury Generation drooled over the big flame that shot the rocket through the sky. *Something important must be happening,* scores of 26-month-olds thought.

We could not take our eyes off the television as the Redstone climbed quickly into a clear pale sky. The *rat-a-tat-tat* of Shepard's static-ridden radio transmissions suggested an overwhelming sense of distance between man and nation: *He's somewhere up there; what's it like?* we thought, laughing, shaking and crying all at the same time.

Was it all real?

Then we decided it was, because miracles happen, as this flight was proving.

But as quickly as the rocket had lifted off, the capsule had separated from booster and was heading back to Earth in a long parabolic arc. Now there were other concerns. Would the heat shield work? Would the parachute open? Would Shepard burn up during reentry?

Pinch us, rouse us from the suspense.

Time now ticked by in milliseconds, a full minute feeling like the lifetime of a planet.

Finally, our fears were allayed. *Freedom 7* landed gently in the Atlantic Ocean, some 302 miles downrange from where it had lifted off.

It was all over.

☆ ☆ ☆

Technologically, Shepard's 15-minute, 22 second flight was little more than a stop between ports of call, and a far cry from Gagarin's single-orbit mission. Schirra, in fact, would later rib Shepard that America's first man in space spent less time in outer space than Schirra took to reenter it.

From a public relations perspective, however, the United States had just won the Super Bowl; we finally had reason to cheer.

We hadn't gotten a foothold in space, yet; that would only come with an orbital flight. But we had a toehold. As Jay Egan would one day remark: "It was like putting your toe in the ocean on a day when you're not sure how cold the water was going to be. That was Shepard's flight."

Even Glenn, who would command the first orbital mission, was surprised at the nation's outpouring of emotion over Shepard's flight. "Most Americans reacted with just an emotional surge that you found almost unbelievable to be in the middle of," he remembered.

This surge of enthusiasm also reached the White House and President John F. Kennedy. Since taking office earlier that year, Kennedy had been a keen supporter of the space program. Project Mercury, he concluded, would revive the nation's morale, as well as his sinking popularity. But the president held no illusions over which nation was still ahead in the space race. After all, the Soviets had already orbited a pilot around the Earth. Kennedy wondered: *What could the United States do to erase the PR and technological chasm that remained between the two countries?*

In a special meeting with Shepard and members of Congress and NASA, Kennedy sought a way in which the U.S. could outdo the Soviets. But what? A trip to Mars? A trip to the Moon? Kennedy had already discussed these options with Lyndon Johnson, head of the National Space Committee, and with von Braun. Von Braun admitted there was a sound engineering basis to suggest the U.S. could develop the boosters and technology to send a man to the Moon and back.

"Kennedy asked, 'Well, what are you guys going to do?'" recalled Shepard. "They said, 'Well, we're thinking about going to the Moon.' That tended to create more excitement in the exploration of space than it did with the competition with the Soviets."

Three weeks later, Kennedy stunned the world with a pronouncement that would forever change the stakes of space exploration. On May 25, 1961, Kennedy told the nation's leaders that it would be the nation's goal by the end of the decade to send a man to the Moon and return him safely to Earth.

Kennedy had issued the ultimate challenge and had given the US the ball to carry. But did we have the technology to realize his dream?

"We thought he was crazy," said Kranz. "But we decided, if that's what he wants to do, so be it. Let's do it."

Making it happen was another matter. The U.S. had amassed a mere 15 minutes of manned flight time. Shepard's suborbital mission had reached an altitude of 115 miles; the Moon was 250,000 miles away. Thus, on the day Kennedy promised the Moon to his country, NASA had completed less than one percent of this ultimate journey.

"We barely understood what it was like to get out of the Earth's atmosphere, let alone into orbit," said Schirra. "We had a lot to learn and I would say it was almost wishful thinking that our technology was that good."

Undaunted, von Braun went to work designing a booster powerful enough to launch a man to the Moon. NASA, meanwhile, continued to march cautiously toward the next hurdle in space exploration: an orbital flight.

In July, 1961, Gus Grissom repeated Shepard's 15-minute suborbital flight. It was a textbook mission until splashdown, when the capsule's newly designed explosive hatch blew off. Grissom nearly drowned, and the capsule, *Liberty Bell 7,* filled with water and sank to the bottom of the Atlantic Ocean.

Although Grissom's flight was a public relations success, its dubious conclusion hardly delighted NASA. As the space agency studied the problem with the hatch, the Soviet Union once again cast cold water on America's dream.

☆ ☆ ☆

If Gagarin, Shepard, and Grissom were space pioneers, Gherman Titov was the world's first true space voyager. On August 6, 1961, Titov, aboard Vostok II, orbited the Earth 17 times during a flight that lasted 25 hours. Everything about this mission was big, compared to the United States' two Mercury missions. Vostok II was 20 feet long and weighed 10,000 pounds. By comparison, Glenn's *Friendship 7* capsule weighed about 3,000 pounds. The Russian spacecraft contained television cameras, radios, and an array of scientific equipment. It also carried enough food, water, oxygen, and control systems items to sustain Titov in space for 10 days.

It was a cruise liner compared to the Mercury tugboat, which could sustain an astronaut for 24 hours.

Unlike Gagarin's clandestine mission, Titov's day-long flight was undertaken without the Soviets' usual veil of secrecy. In the battle of one-upsmanship, the Soviets had once again increased the ante . . . and the United States was headed for another blue funk.

The solution was obvious to NASA: Someone had to climb aboard the treacherous Atlas booster and risk his neck in an orbital flight. Slowly, the spotlight began drifting toward John Glenn.

Glenn was neither a stranger to danger nor to the public spotlight. Of the seven astronauts, he was the most popular and arguably the most celebrated. A veteran of the Korean War and World War II, Glenn had flown 149 combat missions and had been awarded the Distinguished Flying Cross five times. In 1957, he set a transcontinental speed record for the first flight to average supersonic speed ,700 miles per hour.

His strict moral code, due in large part to a Presbyterian upbringing, made him an All-American in the eyes of the public. It was the astronauts' duty to be role models, he argued, sometimes on deaf ears.

Not surprisingly, NASA was not willing to risk the life of an astronaut without further testing of the Atlas. In January, 1961, NASA successfully orbited a Mercury capsule around the Earth. Its occupant? A mannequin. Ten months later, a chimpanzee named Enos circled the planet twice. Enos's mission came just 11 months after Ham, another chimp, had blasted into space aboard a Redstone.

Clearly, Kennedy's goal was as bold as our technological advancements were cautious and small.

But that was with good reason. A manned orbital flight was fraught with danger and questions.

NASA and von Braun understood that any future missions to the Moon could only occur if the astronauts could work freely in a weightless environment for extended periods of time. Some scientists worried that in zero-gravity an astronaut would undergo severe physiological changes. Some hypothesized that his eyeballs would pop out of their sockets without gravity to restrain them. Others feared that in Zero-G an astronaut would become too nauseated to function. Some "experts" even doubted an astronaut's heart would beat regularly during long flights.

That wasn't the worst of it, either. There was the fickle Atlas, a thin-skinned inflatable steel balloon that required pressurization; otherwise, it would collapse and crumble to the ground.

Glenn learned just how unpredictable the Atlas was the first time he and the other astronauts gathered at Cape Canaveral for a nighttime test flight of the booster. The silver Atlas stood shimmering in the coal black night, bathed in pools of searchlights. Stars twinkled above the astronauts' heads, and spirits were high among the NASA engineers and technicians.

The kinks are supposed to have been worked out, Glenn thought, optimistically.

The rocket lifted off in an explosion of sound and fury, turning the coal sky into an immense fireworks display. The speeding rocket stabbed the night with its orange, swordlike tail, painting the black canvas with slashes of color. Glenn and his colleagues, stationed on a camera pad near the launch site, marveled like kids at the booster's power and precision.

Suddenly, something went wrong.

At an altitude of about 27,000 feet, the rocket exploded, right above the astronauts' heads. It looked like an atom bomb detonating in the sky.

NASA would launch five unmanned Atlas flights prior to the liftoff of *Friendship 7.* Two failed. The media was worried. The Mercury Generation was worried.

Jay Egan was worried. *Now this thing is for real. It's just like jumping off a diving board. We don't know if he's going to come back a piece of toast or bacon. Or if a cloud of meteor pellets will tear this thing apart when he goes outside of the Earth envelope.*

☆ ☆ ☆

None were more fearful for Glenn's safety than the townsfolk of New Concord, Ohio. In 1962, New Concord was a close-knit college town of 2,000, and a throwback to an age when small town values—honesty, integrity, concern for one's fellow man—were the way of the land. New Concord, home of John Glenn, astronaut.

Even to tourists, New Concord was a home away from home, perhaps no more so than in the days preceding and following Glenn's flight. Take Route 40 into town, cruise up West Main Street, and visit the town's only diner for a sandwich and a story about John Glenn.

"Hey, pal, even hear how John Glenn and Annie Castor got to know each other? How about an autographed picture of the most popular person ever to live in New Concord? Did you know Bud Glenn lettered in four sports in high school?"

America knew him as *astronaut John Glenn*. Back home, he was better known as "Bud," son of Clara and John Glenn, Sr. And you'd be hard-pressed to find a more popular alum than Bud Glenn at New Concord High.

A regular at Glenn's Plumbing Co., John "Bud" Glenn was a solid B student, who, according to former teacher Ellis Duitch, displayed an interest in politics and science. Glenn was not afraid to put his neck on the line, however, in the name of good old teenage fun. He had a beat-up Chevy coupe, a gift from Dad who once owned a Chevrolet dealership. *The Cruiser,* as it was known to Glenn and his pals. A great way for him and buddy Lloyd White to fly with the wind on a warm summer night. Occasionally, New Concord's version of Huck and Tom opened up *The Cruiser* across the humpbacked bridge that crossed over the railroad tracks. Sometimes they went so fast, the car became airborne.

Fitting, Duitch thought. *It's a miracle they didn't run into someone up there.* Sometimes you just needed luck on your side.

Other times you needed faith and loyalty, as in the Harlequin romance of Bud Glenn and Annie Castor. They had known each other as young children, had dated through grade school and high school, and save for one night, had dated no one else. On that night, Glenn and White teamed on an ill-advised double date (without Annie) that quickly landed Glenn in his sweetheart's doghouse. That was the last time Bud Glenn strayed from his future wife.

To Ellis Duitch, Bud Glenn was more than a solid student, faithful friend, God-fearing Presbyterian. He was a young man who naturally worked for more. On one occasion, Duitch, Glenn, and two other students built a single-tube radio set that could receive transmissions from halfway around the world. Afterwards, this threesome would spend many of their nights at the high school, with the radio's antenna outside, listening to sounds from the other side of the globe.

☆ ☆ ☆

It came as no surprise to Ellis Duitch that John "Bud" Glenn was preparing to be the first American to circle his world. Duitch, however, feared that Bud might not be as lucky as he was with *The Cruiser,* hurdling across the humpbacked bridge.

By February 19, 1962, Glenn's scheduled flight had been postponed nine times. It had happened so often, New Concordians such as 21-year-old Don McKendry wondered if his friend would ever get a shot at making history. In fact, some folks worried that if the hardware and the weather kept failing, Glenn might eventually be too old to fly.

The 2,000 residents of John Glenn's hometown felt about 2,000 different emotions. That was about 1,999 more than their home-grown astronaut felt. Throughout the delays, Glenn remained confident in the rocket, the team of engineers, astronauts and technicians, and, most important, himself.

Like Glenn, Anne Marshall had remained focused on her responsibilities despite the buzz around town. Preparations for Glenn's homecoming parade had consumed her time and much of her emotions. Still, the night before Glenn's re-rescheduled flight, she wondered how the mission would go. She was, after all, making plans for his *return* to Earth. Were these plans premature?

Unlike the more staid Marshall, Don McKendry was in awe—of Glenn, of the flight, of the Moon. *This is my former neighbor who will fly around the Earth once every 90 minutes.* McKendry also was haunted by the possibility of losing the town's hero. Would Glenn go out there and never come back? Would he be forever lost in space?

Duitch retired to his bedroom pondering the odds on Glenn surviving reentry and a water landing.

The night before the heavens would finally welcome astronaut John Glenn, New Concord, Ohio slept fitfully.

Halfway across the country, at Cape Canaveral, John Glenn went to bed at about 7 p.m. He slept soundly, confident in his skills as a pilot, confident in his machine, even if the thousands of components used in the capsule and rocket had been manufactured and assembled by the nation's low-bidders.

At about 1:30 a.m., February 20, 1962, John Glenn awoke, fully refreshed. He showered, shaved, and checked the weather report: a chance for rain. Clouds still hovered over the Florida marshlands, and over Glenn's hopes for success.

The waiting game had begun.

If the many weeks of anticipation weighed on Glenn's mind, his body did not reflect it. During his pre-flight physical, Glenn's blood pressure was 118/80, his pulse rate a steady 68 beats per minute. It was as if he were sitting in his recliner at home watching a soap opera.

Conversely, Jim Oberg's blood pressure was about 300/200 at the moment; Jay Egan would have popped the mercury sphygmomanometer, the instrument used to measure blood pressure.

Some three hours after awakening, Glenn donned the 22-pound, five thousand dollar pressure suit that would protect him in the event the cabin depressurized. Behind the visor of his space helmet, his sharp blue eyes remained steady, focused. He bore the demeanor and appearance of the prototypical heroic voyager: bold, purposeful, stoic, outwardly unaffected.

At 5 a.m., he emerged from Hangar S and strode briskly to the NASA transfer van that would transport him to the launch pad. A crowd of sleepy yet curious reporters watched and cheered as Glenn strode past. NASA technicians bade the astronaut well; one shook Glenn's hand. Finally, Glenn, flanked by surgeon Bill Douglas and Deke Slayton, boarded the van.

The skies remained threatening as Glenn took the 15-minute ride to the launch complex. Mercury Control, meanwhile, had put the flight on hold at T-Minus 120 minutes. After the team arrived at the launch site, Glenn waited inside the van until 5:59 a.m. When he emerged from the van, he was like a movie star attending the opening of a certain Oscar-winning movie. It was almost eerie, Glenn thought, the way the search lights spotted the black sky with streams of milky light; the steaming rocket glittered, almost alive, ready to jump off the pad.

It almost looks staged, Glenn considered.

He cast the rocket a final respectful glance, then boarded the red gantry elevator that would take him to the 11th floor and the waiting capsule. The only question was, would the weather and the myriad switches and relays and motors and meters all hold out this time?

As a pensive Glenn rode up in the elevator, Bob Munkres rose and walked slowly, foggy-eyed, to a room at the far end of his home. It was 6 a.m. and New Concord was just beginning to shake loose the cobwebs of sleep. The dawn of the flight ... as it had been so many other times over the last few months.

Munkres yawned and peered out the window, as he had done each day for the past two weeks, across the street to the home of Clara and John Glenn, Sr. He saw television lights stationed throughout the Glenns' yard, illuminating the shadowy, well-manicured front lawn. The lights glowed brightly, prompting a smile from Munkres. He understood the significance of the illuminated TV camera lights: It meant a news reporter would soon interview Bud Glenn's parents. And that meant the flight was still on. New Concord would soon become the nation's second most popular town, next to Cape Canaveral.

Muskingum College would also be crawling with reporters, thought Munkres, who had taught political science there for two years. Today, America would get a firsthand look at the picturesque school, with its majestic red-bricked buildings, quaint quadrangle. Muskingum College, Bud Glenn's college alma mater.

Munkres dressed. He would watch the flight with his fiancee, Jeanette Fields, head resident at the college's Kelly Hall. Munkres was calm, confident about Glenn's impending flight, and he understood why. He was only a recent addition to New Concord: He had moved to the town from Wisconsin two years earlier, after accepting a teaching post at the college. Also, Munkres had grown up during the Korean War and had seen uncles, cousins, and childhood friends march off to battle. A few didn't return, including a buddy five years his senior.

Glenn was putting his life on the line, Munkres knew. He respected and admired that. But Munkres also recognized that the astronaut had a choice. Munkres's friend lost his power of choice when the government thrust a rifle into his hands. War and death put a different spin on space flight, Munkres decided, ruefully. It brought life back down to Earth.

Nearby, on West Main Street, Anne and Jonathan Marshall also prepared for the flight of *Friendship 7*. Timing, Anne Marshall mused, was as fickle as the weather that had forever postponed Glenn's mission. Today was supposed to be the day Jonathan was to take his college SAT exams. They would have to wait; he was not about to miss this flight. He and his high school classmates had discussed the mission in great detail, and he had shown a growing interest in space exploration.

Jonathan Marshall, however, was not the only person in America putting his plans on hold.

At just before 8 a.m., technicians began the laborious task of bolting on the capsule's explosive hatch. The skies were finally beginning to clear.

At about the same time, Ellis Duitch and countless other dedicated teachers wheeled bulky black and white television sets into their classrooms. Today's curriculum would be briefly interrupted.

Minutes later, Glenn was sealed inside the cramped capsule that would serve as vehicle, living room, kitchen, and bathroom for the next four hours. He was finally alone.

As we watched the first snowy pictures of *Friendship 7* beam across our televisions, we wondered what was going through John Glenn's mind at that moment? Answer: "How would you feel if you were 75 feet atop a missile that might explode?"

Don McKendry would have asked himself the same question were he not engaged in his own hectic mission. Today, the 21-year-old Muskingum College senior was busy at the first of his two part-time jobs, delivering surplus food to the local schools. Of all days, it had to be this day. Glenn had taught McKendry that a person had to make a difference in the world. Maybe delivering meat and cheese wasn't as profound as leaving the planet, but at least McKendry knew he was helping people survive by providing much-needed resources.

Bud Glenn wasn't just a neighbor, he was a friend of the family. McKendry's dad had known Bud all his life. The Glenns were warm, friendly, ethical people. It was easy to be fond of them. Easy to be proud you knew John "Bud" Glenn.

I can't miss this flight, McKendry decided. But how and where would he monitor the flight when he would be shuttling between the food supply station in neighboring Zanesville and every school in New Concord? Borrowing a bit of NASA's ingenuity, McKendry planned to make one of his drop-offs coincide with the launch. He would listen to the updates on radio, then he would sneak into the school gyms. McKendry knew he'd find a TV set there, broadcasting the flight while kids sat spread-eagled on the floor, listening to every word issued.

McKendry jumped into his tattered pickup truck and raced off on his own breathless mission of goodwill. Elementary schools came first.

As McKendry sped off, many of the 1,200 students enrolled at Muskingum College pondered whether they should go to class or follow the flight.

Jay Egan felt his stomach churn. Jim Oberg waited impatiently, as if it were Christmas Day and his wrapped gifts lay tantalizingly before him.

Americans waited, afraid to abandon the snowy images on their TV screens.

Eternity passed, long enough, it seemed, to send a man to the Moon and back.

We were left to our thoughts and fears. As Don McKendry drove down the town's back roads, he pondered the vastness of space, and the opportunity his friend had to be a part of it. Glenn was like Christopher Columbus, visiting the undiscovered world. But Columbus had a choice, McKendry figured. He could always turn around. There was no turning back for Glenn once the mighty Atlas came to life.

Waiting . . . We sighed as a nation.

Finally, Flight Director Chris Kraft consulted with his flight engineers. Were they *GO* or *NO GO* for the flight?

Communications? *GO!* ASCS? *GO, FLIGHT!* AEROMED? *GO!*

Finally, Glenn's acknowledgment: GO!

Then came the countdown.

Ten . . .

Jay Egan strapped himself in for the ride and thought: *A person would have to live in a cave not to be consumed by this flight.*

Nine . . .

Anne and Jonathan Marshall moved their chairs closer to the TV. All thoughts of a homecoming parade vanished from Anne's mind.

Eight . . . Bob Munkres felt his respect for Glenn begin to swell.

Seven . . . Gene Kranz felt the thrill of anticipation melt.

Six . . . Ellis Duitch abandoned his lecture to watch the liftoff.

Five . . . Don McKendry raced into yet another gym of yet another schoolhouse to find the rocket poised for blastoff. He scooted across the floor and crouched with a large group of students seated before a TV set.

Four . . . We began to pray.

Three . . . We held each others' hands.

Two . . . We feared the worst.

One . . . We hoped for the best.

Engine sequence started. Backup pilot Scott Carpenter solemnly bade his training partner farewell: "Godspeed, John Glenn." With a shudder, 360,000 pounds of thrust shook the Atlas to life. Huge pillows of smoke issued from the engine bells, followed by a blinding orange flame.

The powerful Atlas strained and whined in protest as holddown clamps kept the rocket moored to the pad. A second or two later, the vehicle broke loose from its restraints and began its long climb into the baby blue sky.

"We are underway," Glenn reported to Mercury Control, his voice shaking from the tremendous vibration.

Jay Egan watched the rising rocket become smaller and smaller. Every foot the Atlas rose was in itself a tiny miracle, validation of man's ability to triumph over the unknown. *Glenn and the other astronauts are the best and the brightest we have,* thought Egan. *These are mini-miracles they're putting together.*

At 60 seconds, Mercury Control reported that everything was A-OK with the flight. Moments later, the rocket was beyond the view of the television cameras, beyond our visual reality. *Friendship 7* was no longer a part of our world.

Soon astronaut John Glenn and craft were traveling at a dizzying 17,500 miles per hour, 300 miles per minute, four miles per heartbeat. In the time it took to take a deep breath, Glenn had traveled over our home town and four neighboring suburbs.

Up in the tiny capsule, Glenn adjusted to the weightless environment with surprising ease. Working and eating came without difficulty. His eyes didn't bulge out, his heart didn't stop, his breakfast didn't back up. Weightlessness was, Glenn decided, a very pleasant experience.

Glenn's spirits soared similarly when Alan Shepard confirmed the astronaut was "Go" for "at least seven orbits." Those were about the sweetest words the astronaut had heard in a long time. Seven revolutions of the Earth; seven sunrises, seven sunsets, the most reliable weather reports you'd ever get.

"That view is tremendous," Glenn reported back to Mercury Control, with a thrill in his voice. Never had the Indian Ocean looked so breathtaking; never had Earth appeared so fragile. The atmosphere was a thin layer of air, so tenuous, it made Glenn truly appreciate the environment in which he lived.

We could only stare at our TV screens and wonder what our homes looked like from an altitude of 116 miles. Then our imaginations soared into space. *Does outer space actually hold the heavens? Is John Glenn meeting God out there?*

Tom Egan, future pastor of Holy Family of Nazareth Church, considered the flight a means of contacting the Almighty. If God was immense and omniscient, then John Glenn was directly in His presence.

It's unbelievable, thought Tom Egan. It was almost too much to cope with. Such knowledge of God, of the power and potential of man. *Indeed, space holds no boundaries over a man's soul.* Glenn was doing what all men instinctively did: he was quenching his thirst for knowledge.

Many years later, Father Tom Egan would refer to this unique gift of God's knowledge as the "Eighth Wonder of the World."

For Tom Egan's younger brother Jay, the flight of *Friendship 7* was pure magic.

Until something went very wrong.

Shortly after the first orbit, Mercury Control received a telemetry signal that the landing bag in Glenn's capsule might be deployed. If it were, then the reentry heat shield might be loose.

Suddenly, the nation began to worry.

How much heat will he bear during reentry? Ellis Duitch wondered, his worst fear seemingly about to occur. *Can he come back with a loose heat shield? How much protection did the capsule have now?*

Soon, Duitch's questions were answered. *Friendship 7* would be rocked by 3,000 degrees of heat as it reentered the Earth's atmosphere. Without a heat shield in place, the tiny capsule would become a charcoal briquette and Glenn would be incinerated.

Our elation over sticking it to the Soviets had transformed into a chill of dread. We couldn't talk about—couldn't *think* about—the what ifs . . . Jim Oberg's nightmares—had they been premonitions of Glenn's demise? As Don McKendry had once feared, was his buddy about to become a permanent monument to the nation's failed space program?

A hush fell over New Concord, Ohio.

Back up north, in Woonsocket, Rhode Island, Jay Egan, ever the optimist, retained his faith in NASA, in the team.

The team was working to solve this potentially fatal mechanical problem. At length, a potential solution was offered. If Glenn kept the capsule's retro pack on, the metal straps might hold the shield in place. It just might work.

Get home, John, after three orbits, Mercury Control ordered. *Leave the retro pack on.* They offered no explanation; they didn't need to.

During the previous three hours of flight, Glenn's heart rate had averaged 86 beats per minute. It began to climb during a reentry that might have ended his life. During the ionization blackout that precluded communication with the capsule, we held hands, said a prayer together.

Glenn held tight as violent buffeting rocked the capsule like a bucking bronco. His heart rate rose to 109 beats per minute.

The straps that held the retro pack began to burn off in the soaring heat. One snapped off and shot past his window. The capsule was now engulfed in a pocket of fire.

Was that the retro pack being chucked into space? he wondered. *I've got a real fireball here.*

Inside the capsule, Glenn's heart rate continued to climb. Now it was racing 134 beats per minute.

New Concord's pulse made Glenn's appear calm and easy.

When it seemed the nation and the capsule's heat shield could no longer take the strain of reentry, a huge main chute opened to slow the plummeting capsule.

The heat shield had held after all.

When the capsule finally splashed down in the Pacific Ocean, America exhaled as one.

☆ ☆ ☆

From a public relations perspective, Glenn's abbreviated flight was an overwhelming success. Technologically, however, NASA still had work to do to perfect the capsule. During the first orbit of *Friendship 7,* Glenn had lost some of his automatic control stabilizers and was forced to fly the craft manually. Then there was the matter of the heat shield, ultimately no more than a faulty indicator light giving a false signal.

But those of us who weren't NASA employees or design engineers could not have cared less about hardware deficiencies. We were a step closer to the Russians; a few miles nearer to the Moon.

Glenn, a hero before his flight, was practically canonized when he returned home. A quarter of a million Washingtonians welcomed him home during a stirring reception in D.C. But this was a mere surprise party among a few close friends compared to the reception the astronaut received in New York City. An estimated four million people turned out in lower Manhattan and deluged Glenn and the streets with nearly 3.5 tons of ticker tape in one of the largest and most impassioned celebrations of its kind.

But these festive processions honored John Glenn, *astronaut.* On March 3, Anne Marshall and the New Concord school board paid homage to their neighbor and friend, Bud Glenn.

The weather was chilly, the feelings warm. On the day Bud Glenn returned home, like a beloved brother returning from battle, New Concord's population exploded from 2,000 to almost 50,000. A potential windfall for the town's Dairy Queen, which already served up its share of Glenn burgers and Annie hot dogs.

Over the previous 24 hours, tourists had poured into town in preparation for Glenn's homecoming parade. Now, as the reception began, locals and the out-of-towners jockeyed for the best parking spots and vantage points along Main Street, with the same determination as Mercury astronauts vying for the first ride in space.

When the parade began, people stood as many as eight rows deep, in front of the grade school, the doctor's office, the funeral home, the post office, and through the center of town.

Kids hung upside down from trees to get a glimpse of Bud and Annie Glenn passing by in a sedan convertible. Much of the media was utterly unprepared to deal with this throng of humanity; professional newspaper photographers battled with dozens of amateur photographers for the opportunity to snap a single publishable picture of Glenn.

Other members of the press were a bit more ingenious, however. For them, Duff's Sunoco was becoming the headquarters from which they would launch their assault upon Glenn with cameras at the ready, sightlines unobstructed by even the tallest of the town's maples and oaks.

They owed their success to a 21-year-old attendant with a knack for solving problems. Six years later, this enterprising young man would become principal of John Glenn High School. His name: Don McKendry.

On this day, McKendry was pumping gas into a zillion cars instead of pumping hundreds of pounds of surplus food into school cafeterias. Before the start of the parade, McKendry figured he was about to miss another historic event. But his fortunes changed when the first few enterprising reporters made an offer McKendry could not refuse.

"I'll give you twenty bucks if you let me borrow that ladder," a determined newsman offered. McKendry could have used a few more hands and a couple more gas pumps, but a ladder he didn't need. He did, however, recognize that he held the upper hand in this business deal. Clearly, the photographer knew he couldn't get a serviceable shot of Glenn if he couldn't see over the crowd. The ladder would eliminate that barrier.

"Twenty dollars? You got yourself a deal." And so Don McKendry, student, delivery man, gas station attendant, became a businessman. One by one, cameramen approached McKendry, $20 in hand. They left a few dollars lighter, but guaranteed a bird's eye view of their subject. By day's end, Don McKendry's pockets were stuffed with dollar bills.

New Concord, Ohio had emotionally been swept away on a bed of confetti and ticker tape. Doubtless, John Herschel "Bud" Glenn was the most famous person in town, maybe in the nation. His flight had infused America with a new sense of pride, and it had helped erase the humiliation of Gagarin's flight and the Bay of Pigs debacle.

☆ ☆ ☆

Man's appetite for space exploration was just beginning to grow, and the Moon still glowed in our bedroom windows at night.

For Deke Slayton, a closer view of our nearest celestial neighbor would have to wait. An irregular heartbeat scrubbed his flight, which was originally scheduled to follow that of *Friendship 7*. Thirteen years would pass before Slayton was declared flight-worthy. In 1975, he joined Tom Stafford and Vance Brand on the historic Apollo-Soyuz joint mission.

Scott Carpenter's *Aurora 7* and Wally Schirra's *Sigma 7* followed; each flight was more challenging, each longer in duration. In May, 1963, Gordon Cooper and *Faith 7* concluded Project Mercury on a promising note. Not only was he the first American to fly in space for over a day, but when his automatic controls failed, Cooper proved that an astronaut could, without question, fly a spacecraft.

Project Mercury confirmed that America could send a man into space and bring him home safely. It proved that a man could function with little difficulty or discomfort in a weightless environment. America was still a step behind the Soviets, but that would soon change with projects Gemini and Apollo.

Most important, Project Mercury gave Jay Egan, Jim Oberg, Don McKendry and all the members of the Mercury Generation a sense of hope, a wide-eyed way of looking at life. *We flew into space, Wow! Maybe we can do anything.*

Project Mercury encouraged us to fashion capsules out of plain refrigerator boxes; it would later inspire us to fly model rockets in fields, pretending we were Wernher von Braun. In the years to come, the flights of the Mercury astronauts would move us to head down to the Cape, and experience the power and beauty of a Space Shuttle liftoff.

Mercury let us know that it's okay to dream.

Liftoff

Nine-month-old Daniel Jutras slept soundly, snuggled in the arms of his dad, 36-year-old George Jutras. Daniel dreamed not of astronauts marooned in space nor of Redstone rockets exploding on the launch pad at Cape Canaveral. His were the blissful dreams of a child secure in a world of unconditional love; he was too young to know that in minutes he would be thrust into a different world that had given Jim Oberg, Jay Egan, and Don McKendry nightmares of space men eternally lost in space.

George Jutras, a veterinarian by trade and aficionado of space exploration by hobby, gazed lovingly at his son. This was a nightly tradition in the Jutras household: First, dad fed Daniel his bedtime bottle; then the two cuddled happily until the handsome boy with the blond hair and blue eyes drifted slowly, contentedly to sleep.

This is our quality time, George Jutras reflected. *It helps heal ulcers just to hold a baby.*

Jutras began to drift mentally, and the hypnotic drone of the television was a sedative for him. He had spent a long but fulfilling day at the veterinary hospital, caring for black Labrador retrievers, parakeets, and Siamese cats. Now he needed to unwind.

Suddenly, an excited voice on the TV roused him from his reverie. . . . *Liftoff of the Space Shuttle* Atlantis.

That's what I forgot, Jutras winced. With a snap of his head, he was alert. Earlier in the week, he had decided to share this liftoff with his son. Although Jutras had witnessed some 15 launches in person, Daniel would be seeing his first. Smiling, Jutras recalled his first live launch: *Columbia,* 1981. The maiden voyage of the Space Shuttle. He sighed wistfully.

Today's launch was special for another reason: It was a night launch. From the Jutras's home in Sebastian, Florida, 40 miles from Cape Canaveral, night launches were fireworks displays in their backyard, only NASA used nuclear firecrackers. Jutras was accustomed to awakening at 2 a.m. for these nocturnal extravaganzas. Tonight's launch was scheduled for 10:34, prime time; fortunately, Daniel would not lose much sleep.

Dad scooped up son, and together they ventured outside.

They chose an unobstructed view on the lawn, by the in-ground swimming pool. Daniel was wide awake now, thanks to the enthusiastic greeting from the family dogs: Annie, a tan mixed Labrador; Cody, a rust-colored Doberman; Harry, a salted Chihuahua/dachshund mix; and Disney, a three-legged sheltie. All the dogs were animal rescue cases, and at least three of them would not be alive were it not for George Jutras's benevolence. Daniel giggled back at his friends, and gave them a hearty wave.

With that acknowledgment, the dogs relaxed, and dad and son began their space vigil.

Heavenward, a full moon was a huge spotlight in the black Florida sky, and it illuminated the canal in the backyard. It was a perfect night to watch a rocket leave the planet. For Jutras, each Shuttle voyage was a Space Mountain ride at Walt Disney World; each was an adventure that left him gasping for more and wondering if he could even bear more. He had been intrigued with space exploration since the days of Apollo; had watched the missions on television, had even built models of Apollo XIII as had many kids.

Tonight was dedicated to introducing a new generation to space flight. Jutras just hoped his son was mature enough to focus on the event.

The minutes ticked off, then the seconds. Finally, the countdown began. At T-Minus three seconds, Jutras put the safety of the crew of STS-86 into the hands of a higher source.

Dear God, be with them.

Two, One, liftoff!

A combined thrust of 5.2 million pounds shot the craft off the launch pad. A brilliant orange-white flame speared the stark sky, transforming the scattered shadowy purple clouds into fluorescent orange pillows on a black blanket of night. *It's like watching fireworks amplified one hundred times,* an awestruck George Jutras thought, grinning.

Dad squeezed son as the mighty rocket zoomed out of their world. Jutras gazed into Daniel's face, and the little boy's eyes were as alive as the black, violet, gold, and orange sky. His face was a mask of wonder and delight.

This is so great to share, Jutras thought. *Daniel might not remember it later, but for this moment we can look at each other and share this.*

When *Atlantis* had finally disappeared from view, Jutras carried Daniel back to their five-year-old colonial home. He held his starstruck son against his chest, and together they watched the remaining minutes of CNN's coverage of the flight. The network's powerful zoom lenses afforded viewers an opportunity to see the Shuttle zoom toward the verge of outer space. About nine minutes after liftoff, the spacecraft and its crew would be in low Earth orbit.

A mission is like my college education, Jutras considered, as he caressed his son's head. Four years of undergraduate work, another four years of veterinary school. Years of work to become a doctor; years of work to become an astronaut, to launch a career, to launch a mission. The launch itself was analogous to the time he took his Veterinary National Board Exam: if he passed the crucial tests, he was on his way. Same with this liftoff of STS-86: if the astronauts and the rocket passed the test, then the crew was on its way. Jutras smiled to himself as he recalled the feeling of relief when he learned he had passed the Boards: *I'm gonna be a veterinarian.*

That's exactly the way the people on the ground must feel when the spacecraft goes into orbit: they've done it right. What a relief.

Some six minutes after the first flames spat out of *Atlantis's* engine bells, three minutes before the vehicle would enter into a low Earth orbit, Jutras felt the couch begin to shake. Then the house began to rumble with a low roar. The sound waves of the liftoff had finally reached his home. He glanced into the dining room, to a row of decorative Norman Rockwell plates mounted on a strip of wood on the wall. The plates tapped, then jingled as the tremors clattered them together.

This was the final ear-shattering, ground-rocking boom heralding the culmination of this celestial fireworks display.

Just as suddenly, the ground was still, the house silent, save for the chatter of the television commentator. The journey of STS-86 and crew was just beginning; similarly, for George and Daniel Jutras, their voyage together into space was just starting to unfold.

☆ ☆ ☆

It was as dark as a coal mine. The stark blackness crept first into my mind, then into my senses. My eyes were already closed, but I squeezed them tighter, for they felt as puffy and swollen as those of a boxer who'd taken one jab too many. I had not slept in nearly 48 hours, and my mind and body were frazzled with exhaustion.

I breathed deeply to calm my racing heart. A relaxation technique. I'd learned it the year before, during my junior year in college. *Feel the fingers relax, the back, the legs, eyes. Relax the mind*, I told myself. Another deep breath, exhale slowly.

Still, as my body numbed, my mind spun with myriad thoughts.

I can't believe I'm here. Was it fate?

There was no sound, no sense of smell, no sight; near sensory deprivation. I felt like a Mercury astronaut bugging out in a NASA isolation chamber.

Slowly, a floating feeling began to warm me. I sensed no front, no back, no top, no bottom. Utter disorientation.

I'm weightless.

Suddenly, I began to tilt. *I'm falling!*

I forced my eyes open and regained my balance.

I would not fare well in a Zero-G environment, I decided.

I was not in a weightless environment right now, it just seemed that way. Fatigue, the smothering darkness, and a handful of beers ingested hours earlier had made me feel spacey, an appropriate feeling, considering my present point of reference: I was sitting on the roof of my buddy George Jutras's red, '69 Chevy. We were parked just off Route A1A, in Titusville, Florida, in a nearly deserted marsh, across a lagoon from Cape Canaveral, where the Space Shuttle *Columbia* was a glowing white candle, tossing a bulb of light against a blanket of coal-black.

STS-1, the mission that would usher in a new era of space exploration and research.

I had no clue whether I was three or seven miles from Complex 39-A, the launch pad. Which of the area's many marshlands and lagoons had we happened across? Who knew? I didn't even know if it was legal to park here. Perhaps an alligator would creep through the waist-high grass and bite me on the big toe.

The exact longitude and latitude of this exotic world was inconsequential. I only cared that we were here. Just across the water from a real spacecraft!

I jumped to the ground, then peeked inside the car at the glowing dashboard clock: *Three a.m.* We had been here just 30 minutes, but it seemed like a lifetime. To see the Shuttle so close, so real, flooded me with memories of my youth, when I watched the Mercury astronauts cheat death in flying tin cans. Ironically, today, April 12,1981, marked the 20th anniversary of Yuri Gagarin's historic inaugural mission: the voyage that fueled the U.S. and Soviet Union's desperate race to own the heavens.

I looked deeper into the car and saw the shadowy figures of my four friends. They were in various stages of unconsciousness. *Do you guys understand the significance of this flight? Have you even heard of Alan Shepard?*

Of the five of us, I was the eldest statesman—age 22, a college senior, the only one old enough to recall the dramatics of a Redstone liftoff. Greg was a sophomore who hailed from one of the many preppy Rhode Island towns. The University of Rhode Island, which we attended, was a pastoral catwalk for fashion plates such as Greg. That is, he wore nothing but triple-decker IZOD polo shirts with collars that defied gravity, scuffed top-siders, and two-tone pastel pants, splashed with tiny porpoises or other such fish and mammals. Even though he was two years my junior, he was the most vocal of the group, the decision maker.

Then there were Ali and Monie, our Iranian friends. Ali was about my height, 5'6", just the right height to stare at women's breasts, which was his favorite pastime. Ali had been born with an erection that, to the best of my knowledge, had never been relieved. Consequently, his desperate black eyes were always wide and vacillating from 38Ds to 36Bs, like some form of radar.

Monie was too busy staring at his own reflection in wall-sized mirrors to worry about sex. He was also some five-and-a-half feet tall, but was a miniature mountain of a man. He lifted weights obsessively, and the campus rumor had him bench-pressing a baby elephant every Tuesday afternoon. Monie owned a Browning rifle that could fell a herbivorous mammal grazing some 120 miles away in Fall River, Massachusetts. Monie even had his own bullet-making machine. Although frighteningly intense, he possessed an easy sense of humor that could be sparked at the unlikeliest of times. This was fortunate for the rest of us, for we didn't care to have our testicles shot off by a psychotic hunter.

Lastly, there was George Jutras, my one close friend in this clique. He and I had lived in the same dormitory for two years, and we had struck a quick and lasting friendship. We had much in common: We were both short and obnoxious, although I was considerably more haughty than he. We were both intensely driven: I was studying to be a journalist, he was banging away 25 hours a day to be a veterinarian.

I looked for signs of life from the guys, saw none, and withdrew from the front passenger's window and pulled myself back onto the roof of the car. T-Minus four hours and counting. There was nothing to do but wait and daydream. Quickly my mind drifted back a few days, to the day our travel plans had begun to take shape.

☆ ☆ ☆

Four p.m., a rainy Tuesday afternoon. I rocked back and forth on a spindly-legged bar stool, cup of warm beer in hand. I was in the back room of a packed URI Pub, my favorite chemistry lab between the hours of dawn and dawn. A beer-splashed television, with faulty vertical hold, hung from a rusty nail on the far wall. Images of destruction, drugs, malice, menace, jay-walking, murder, suicide, and death assaulted my senses. A news brief was on. Annoyed with the wretchedness before me, I took a deep breath; the stale/sour smell of dried beer and vomit choked me. I held my breath. Suddenly, the air exploded out of my lungs, from hearing the exciting announcement that the Space Shuttle *Columbia* would lift off this Friday.

Friday. Too bad, I thought. We were leaving for Ft. Lauderdale that same evening, to do what college guys do best: drink beer, drink broads. Could we rearrange our schedules in order to leave Wednesday? We'd get to the Cape in plenty of time to see the rocket lift off. *Imagine seeing a rocket lift off in person; how magical that would be!*

Maybe the alcohol was toying with my emotions, or maybe the thought of witnessing a live launch had revived the little kid in me. Either way, I was thinking about space exploration for the first time in years.

It had been six years since Americans had ventured into space. That came with the United State's historic joint mission with the Soviets. With Apollo-Soyuz, the U.S.-Soviet space race had come full circle: Our bitter enemies on the ground had become our compatriots in space. Unfortunately, the public's interest in space exploration had long since waned.

From 1975 through 1981, NASA moved ahead with the Space Shuttle program. If a reusable rocket was more economical for the government, it was less appealing and intriguing to the nation's devout fans of space exploration. On one hand, this was just a plane that took off vertically and glided back to Earth. No more, no less. Somehow, anonymous astronauts doing research in Earth orbit did not hold the same charm and excitement as one brave man crammed into a pod, headed to the unknown.

On the other hand, I also recognized that the days of Mercury, Gemini, and Apollo were well behind us. I couldn't live in the past forever, could I? Skepticism and apathy notwithstanding, in my heart, I knew I'd jump at the chance to witness a Shuttle launch.

Alas, I knew there was no way to work around our course schedules. We were students, after all.

☆ ☆ ☆

At 7 p.m., Friday, April 10, we piled into Jutras's dilapidated Chevy. John Glenn's *Friendship 7* was a mobile home compared to this coffin with hubcaps. We were crowded two in front, three in back, and would be stapled together for the next 24 hours. 1200 miles. No stops, no food, no sleep, no urine evacuation tubes. This was worse than John Glenn's *Friendship 7* mission. Welcome to the mission of *Friendless 7*.

With a well-oiled shoehorn, we squeezed ourselves into the roadcraft. Jutras popped *The Best of the Doors* cassette into the car stereo; this would be our sole source of entertainment for the next day. Little did we know that, as we drove through the darkened, quiet campus, the scheduled liftoff of flight STS-1 had already been postponed. A computer glitch had set the liftoff date back two days, to April 12.

Our lives were about to be changed forever.

☆ ☆ ☆

And so we drove and drove and drove. It would have been quicker and more enjoyable to drive to the Moon. To say we were roughing it was tantamount to saying Wernher von Braun liked to play with toys. *Friendless 7* got three miles to a gallon of gas and three city blocks to a quart of oil. The engine was like an old man: constantly wheezing, perpetually leaking all over himself.

The stereo? The crooners of the 1920s produced sweeter, clearer sound barking into a plastic megaphone. Jutras's radio didn't work, but at least we had that one Doors tape to listen to. More than 24 hours of Jim Morrison belting out *Touch Me*. Nearly 1,500 minutes of Monie and Ali harmonizing in thick, Middle Eastern accents:

Come on, come on and touch me, babeeeee. Cannot you see zat I am not afwraid? Vat vas zat promise zat you made? Vy don't you tell me vat she saiiid? Vat vas zat promise zat you made?

By the time we reached Orlando, some 27 hours after the start of our mission, I prayed the cassette would end up in a huge fire, a funeral pyre, to be exact.

We pulled into a nondescript economy motel and all five of us packed into one room. All told, we had collectively slept about eight minutes, and we smelled like Frank Borman after eight days in space. Thus, neurologically, we required many beers and much more sleep. But first, we decided to take a dip in the motel's pool. We unpacked, changed into swim suits, and popped open a beer apiece. I turned on the television. I had gone a full day without TV and had felt like a newborn child without its mother's breast. I needed my mental nourishment, even if it was, again, the Godawful news. The newscaster was one of those barely pubescent twits whose eyes visibly moved from left to right across the screen as she read a script from the TelePrompTer.

"Now, an update on the Space Shuttle *Columbia* . . ." she said, nearly mispronouncing the word Shuttle.

"Quiet, guys, I want to hear this," I shouted at the others, who were murmuring among themselves about tits and asses.

"After being postponed due to a computer malfunction, the Shuttle will lift off tomorrow morning at 7 a.m."

Tomorrow. TOMORROW!

I looked at my watch; it was now 11: 08 p.m. I searched for Jutras, and our eyes met.

"This is too important to miss," said Jutras. "We have to see this."

"Where are we and how do we get there?" I asked, illogically. I was getting excited. Could it be? Nope, I wasn't even going to entertain the thought. . . . A live launch?

Jutras rummaged through his tattered *faux* leather suitcase and pulled out a map of Florida. We all crowded around him, fingers poking at the map.

"We're about 60 miles away from the Cape," declared Jutras. "If we can get to Route A1A, it will take us right to the Cape."

Suddenly, time stood still, and we all froze. Decision time: Beer and broads for breakfast, or a Shuttle liftoff?

"Let's gooooooooooo!," we screamed in unison. Then clothes, six-packs, hair dryers, and condoms flew all over the room and miraculously back into the suitcases and duffel bags. By the time we relieved our severely stretched bladders, reloaded the car, and checked out of the motel, it was 12:30 a.m. We piled back into *Friendless 7* and cranked the cassette player all the way.

Somehow, we managed to arrive at the stark, deserted marsh where we would become a part of history.

☆ ☆ ☆

I cleared the mental cobwebs with a shake of my head and shelved this mental retrospective with the other volumes of my moderately interesting life.

Sitting atop a car lends itself to musing, and I was considering the similarities between the flights of *Columbia* and *Friendship 7*. Both missions were fraught with unknowns. The Shuttle was a new machine, whose hardware and solid rocket boosters had not been tested in space. Similarly, the Atlas booster had been erratic and a source of concern for NASA, until Glenn confirmed its flight-worthiness during the flight of *Friendship 7*. NASA engineers and scientists had wondered if Glenn could survive weightlessness. Now, I wondered what would happen if the Shuttle exploded directly over our heads.

The political climate of the nation at the outset of the Shuttle program mirrored that of the Project Mercury era. Iran had replaced the Soviet Union as the nation's primary enemy. Three months earlier, a group of Iranian militant students had released the remainder of the 66 U.S. embassy envoys it had taken hostage November 4, 1979.

Now we needed another Alan Shepard or John Glenn to help the nation feel good about itself. Today's heroes were John Young and Bob Crippen. Young was one of the agency's most seasoned pilots, a veteran of two Gemini and two Apollo missions. Crippen was a rookie, who had worked on the Skylab program and Apollo/Soyuz, and was involved in design aspects of the Shuttle.

I smiled. *You and I share a common bond, Mr. Bob Crippen. We're both rookies, and we've waited a long time to be a part of this mission.*

I wondered how Crippen felt at this moment.

☆ ☆ ☆

At 4 a.m., astronauts John Young and Bob Crippen rode the transfer astrovan to Launch Complex 39-A. After the first scrubbed mission, Crippen was excited but apprehensive. He knew the Shuttle was a complex vehicle; consequently, he feared this flight might be postponed more than once, as NASA's engineers and technicians worked out the bugs in the boosters and hardware.

Still, he had complete confidence in the machine and the program. The powerful boosters had been tested and re-tested again and again on the ground; the many on-board computers had worked to near perfection. Even the anomaly that had scrubbed the mission two days earlier had been quickly and easily resolved. The thousands of tiles required to protect the Shuttle and its crew during reentry now adhered perfectly to the spacecraft.

For Crippen, today was the culmination of 12 years of hard work and immense patience. When he entered the astronaut corps, Deke Slayton, then the head of the astronaut's office, had warned Crippen there would be no flight openings until the Shuttle program. Crippen didn't mind. He felt tremendous satisfaction contributing to the program in other ways, and he found his work to be immensely rewarding. This flight was the whipped cream and the cherry on top of it; a great way to cap off that work.

Crippen gazed appreciatively at the Shuttle, bathed in shimmering light. It was overwhelming, a living, breathing kind of beast. As it sat there, he saw that it had a unique kind of beauty, although it was not streamlined like its pencil-shaped predecessors.

Some three hours before he would experience his first liftoff, Crippen joined Young in the gray launch service elevator. They would ride it up to the White Room that was connected to the open hatch of the

Shuttle. Then they would be loaded, Commander Young first, Pilot Crippen second, into the orbiter's seats. Straps would be tightened, buckles secured, legs and feet secured. The astronauts would be tipped over backwards, as if in a living room recliner.

It was time to focus on the mission.

☆ ☆ ☆

As time passed, other vehicles pulled off the road and joined us in the marsh. By 5 a.m., our secluded spot was teeming with people, like the infield at Churchill Downs minutes before the running of the Kentucky Derby. The once silent swampland was buzzing with excitement, and a nervous anticipation had replaced exhaustion and *ennui.*

This is really going to happen! It isn't a dream!

Car radios hummed in the night, blaring music into the ceilingless sky. Portable lights and car lights illuminated the marsh and I could see people breaking out cameras, movie recorders, American flags, and, of course, a six-pack or two.

To think that a few days ago, I could only dream of being here. It all seemed surreal now, as if fate had drawn all of us to our personal launch site. Had it not been for one faulty computer, I'd likely be sitting in a Fort Lauderdale hotel room, envisioning nubile coeds in tiny, string bikinis. I chuckled at the irony. Suddenly, my mind was flooded with the memory of mom's sticky turkey baster, my first Mercury Capsule. The ratty hooded coat, my pressure suit. Dad's Fairlane 500, my Redstone. Grandpa's battered, gold recliner, my custom couch. My personal journey into space had come full circle now.

It seemed ironic that 35 years ago space exploration was relegated to fodder for the pulp fiction and science-fiction magazines. The exploits of the Mercury astronauts had transformed sci-fi into science fact. Would the flight of *Columbia* affect the nation the way *Friendship 7* had? Would it spawn a Shuttle Generation?

☆ ☆ ☆

By 5:30, the sky had turned a murky pale blue, as day struggled to eclipse night. The lagoon changed from dull violet to pale gray. The marsh actually looked less like a swamp and more like a meadow, complete with tall, oat-colored grass waving in a gentle breeze. With the huge MGM spotlights quashed, the Shuttle was a giant white tablet. It looked so small, so powerless from here. What would liftoff be like, I wondered? The Redstones popped off the pad like an Estes model rocket, and the three-story Saturns clutched desperately at the sky as they inched their way toward space.

Would *Columbia* climb directly over our heads?

☆ ☆ ☆

Bob Crippen was living by the astronaut's creed: Focus and don't screw up. Although he had experienced liftoffs only in training simulators, Crippen intimately knew the sequence of events that would precede the launch. The first dramatic step would occur with the ignition of the main engines. For about six seconds, they would burn furiously, then things would happen very quickly, very powerfully.

☆ ☆ ☆

Six thirty. We were five grooms at the altar, wondering if the bride would make it safely to the church and on time. Jutras looked at me, his weary munchkin face lined with concern. "I hope nothing goes wrong; I'd hate to see this postponed now," he said. Jutras was right. We'd come too far to leave *Columbia* standing on pad 39-A.

I wonder if the astronauts have their fingers crossed right about now? I wondered.

The air was thick with tension, and people's faces were taut with worry and excitement. Looking to my left, I saw a balding man, corpulent, fortyish, sitting hypnotized in a ramshackle lawn chair. The man stared wide-eyed at a snowy picture on his tiny, portable TV. Walter Cronkite's stern yet friendly visage filled the screen, but his voiced was muffled amidst the hubbub of the marsh. I didn't need to hear, however. I knew from watching so many previous launches that Cronkite was detailing the crew's pre-launch activities.

Panning to my left and then behind me, I spotted rows of cars, trucks, and campers as far as the eye could see. There had to be thousands of curious people here, many antsy and pacing nervously before their vehicles. Others sat on the roofs of their cars and trucks and stared pensively across the water.

There, the Shuttle was the silent child preparing to come to life.

☆ ☆ ☆

At T-Minus 30 minutes, technicians in the White Room and the ground crew left the service tower at Pad 39-A. They relocated to a safe position three miles from the launch site. Only the crew remained, safe inside the delta-winged glider. They were very much alone now.

☆ ☆ ☆

At about the same moment, Jutras reached into his duffel bag and pulled out an archaic, clumsy, eight-millimeter movie camera. He had brought it, ostensibly, to film our trip to Ft. Lauderdale. Thank God he had it; to capture this historic launch on film would be remarkable. What a gift to bestow upon our future children!

☆ ☆ ☆

T-Minus seven minutes. The White Room and catwalk swung away from the orbiter.

T-Minus five minutes. The orbiter switched from ground power system to on-board electrical system. Liquid oxygen and hydrogen fuels began to pressurize. The crew lowered their helmet visors and turned the audio volume to high *in their headsets.*

☆ ☆ ☆

Five minutes. Thousands of nervous stomachs rolled simultaneously. A prep, two runts and two Iranians perched themselves upon the roof of the Chevy. We glanced at one another reassuringly, as if we were the astronauts about to bid farewell to our home world.

☆ ☆ ☆

T-*Minus :60. Crippen glanced over to Young. "Hey, we might really do this," he remarked to his commander. Crippen's heartbeat was 130 beats per minute, Young's a calm 90 beats per minute. Young, the steady vet, could not have increased his heart rate even if he wanted to. For the rookie flyer, however, the excitement was building to a crescendo.*

One minute to go. We leaped off the Chevy and stood side by side at attention. No one blinked in this swelling, hushed crowd of astronaut wannabes.

T-*Minus 30 seconds. Management of the liftoff switched from Launch Control to the Shuttle.*

The thousands of spectators on hand took a collective deep breath. A voice from a nearby portable television heralded the countdown: *Ten, nine . . .*

At T-Minus :08, thousands of gallons of water were released into the base of the launch pad to help dampen the ear-splitting roar of liftoff.

Seven, Six . . .

T-Minus :05. The on-board computers sent a signal that opened a valve, releasing liquid oxygen and hydrogen to the orbiter's three engines. The liquid fuels were converted into gasses, compressed, mixed and ignited to produce 1.1 million pounds of thrust in the engines.

Four, three . . .

The solid rocket boosters ignited and combined with the main engines to produce a total thrust of 5.2 million pounds. The boosters began to crackle with life; the vehicle strained against hold-down clamps, which kept it moored to the tower. Even though Crippen and Young wore helmets, the tremendous noise and vibration rattled in their heads.

One, liftoff!

Columbia strained to break free from the metal shackles that held it Earthbound. Bright orange flames flashed from the Shuttle's engine bells, and huge plumes of white smoke filled the air. The Shuttle cried for life and finally tore free from its restraining clamps.

The rocket leaped into the air, hurdling the tower in a burst. My eyes could hardly keep up with its lightening-quick ascension into the clear blue sky.

For Crippen, liftoff went as he'd expected. It was not a violent affair; more like driving a pickup truck along an old country road. The blare of the solid rockets igniting tore into Crippen's gut. It shook him, made him unsure what was a physical sensation and what was an emotional feeling.

The rookie pilot monitored the liftoff, ready to follow abort procedures if necessary.

I had waited nearly a lifetime for this moment, and now it was happening so quickly! One moment the Shuttle stood idly on the pad; now it was soaring and yawing directly over our heads, flying like a dream. Beside me, Jutras had the movie camera trained on the flash of white and orange that was speeding out of sight. A quick glance around: Thousands of mouths agape. Not a blink to be found in the enormous crowd.

Suddenly, the ground began to rumble, and a tremendous roar of sound rolled across the marsh. The shock waves from the liftoff had finally reached us. I supported myself against the car until the earthquake-like tremor ceased. The Shuttle was no more than a white dot dabbed against a background of robin's egg blue.

Then the mighty rocket was out of sight. The ground was still, the marshlands eerily quiet. A few wisps of smoke floated and curled in the blue sky.

All at once, the crowd erupted in unabashed celebration. A middle aged woman, with long, golden hair, wearing a dusty brown summer dress and ostentatious hoop earrings, began to dance with joy. People hugged and kissed their neighbors, as if in the midst of Sunday Mass. Adults-turned-children hurdled rows of cars to shake hands with people who bore grins broader than the Moon. American flags were unfurled, waved, and tossed from person to person. At least 36,000 high fives were slapped.

We cried.

Ali, Monie, Greg and Jutras were a blur through a glaze of my own tears. I wiped my eyes with a shirtsleeve and composed myself. I climbed back onto the Chevy and enjoyed the celebration. It reminded me of the day one year earlier when the U.S. Olympic hockey team beat the Russians. *Do you believe in miracles?* Yes!!!! A collection of fuzzy-faced, gap-toothed upstarts from frozen ponds around the country had brought us together. That miracle was happening again today.

Why did it have to end?

For many of the others here, the party was just beginning. Corks popped off champagne bottles; beer cans opened with a *fwitz.* I should have joined in the festivities; but all I could do was gaze numbly into the sky. *Come back, guys. I wanna do it again.*

Suddenly, I felt a hand on my shoulder. It was Jutras, all bubble-eyed, grinning dopily. I permitted myself a smile. God, I was tired.

☆ ☆ ☆

Reality struck us at about the time *Columbia* was in its third orbit: It was time to get back to life. The real world included real cars and real traffic and hundreds of vehicles rolling and yawing a mile an hour up Route A1A. We boarded *Friendless 7,* Jutras commander, myself, the navigator.

"I hope we have enough oil to get us to Fort Lauderdale," a low, scratchy voice mumbled from the back seat.

☆ ☆ ☆

Two weeks later, Jutras and I stood in our bathroom at URI's Dorr Hall. He and I lived in a dormitory suite composed of a small central lounge, four bedrooms, and two bathrooms. It was an economy motel for eight college students. The bathrooms were comfortable, and they included a bathtub with shower, vanities, and a private toilet. This setup afforded us a measure of privacy foreign to most dormitory bathrooms, which housed toilets for 20 residents. Still, a private privy was not an ideal place to view a movie with a dozen friends. It was, however, as dark as the Titusville sky at 3 a.m., and the pale lime-colored walls served adequately as movie screens. Tonight's feature: The launch of STS-1, *Columbia.*

This was the whipped cream and cherry of *my* life-changing experience. For me, the power, the magnitude of the launch of *Columbia* grew in splendor with each mile we put between ourselves and that Titusville marsh. *Had it happened?* I asked myself during our return ride home. Yes it had. This recording would forever keep it real.

Jutras, in particular, had been profoundly moved by the launch of *Columbia.* "Florida seems so magical to me now," he told me afterwards.

He was right. Even now I felt as though I still had magic dust on the soles of my sneakers. As a student studying and practicing to become a professional journalist, I thrived, swooned, laughed, cried, and gritted my teeth every time I set word to paper. Watching the Shuttle blast off confirmed for me the power of man. Perhaps one day I would use words and images to shape the way people thought. Same with Jutras: One day, he would be the attending physician at a veterinary hospital. I just knew it.

And he and I would always be bound by the voyages of *Friendless 7* and *Columbia.*

☆ ☆ ☆

Within minutes, our makeshift theater was chock-full of wide-eyed fans of space exploration. I had a bird's-eye view from the back of the tub, which served as a balcony for those of us who were height-challenged.

The lights went off, and I recalled that delicious feeling of nothingness. Mentally, I was back on the roof of the Chevy, feeling as weightless as John Glenn.

The projector flicked on, and the first images of the historic flight flashed back at us: the pod-like Shuttle on the horizon. The swelling crowd. Thousands of furrowed brows, faces tighter than a snare drum.

I panned the dimly lit room and saw looks of amazement.

Then came the liftoff, the Shuttle streaking into the sky, almost faster than Jutras was able to record its path.

"Wow!" the crowd of 10 exclaimed.

"I can't believe you guys were actually there. You are so lucky," said another voice.

Then I saw myself on screen, but it was not *me*. It was that little boy who had marveled over all the Mercury, Gemini, and Apollo missions. How surreal it was to watch myself follow the Shuttle into the sky. Now, as an observer, I could see that we were all a part of the mission, as was every person who witnesses a live launch.

This really didn't happen, I thought. *I'm living one of those bizarre lucid dreams.*

If it were a dream, then we were all asleep.

Finally, sadly, the film was over, and the lights went back on. Reluctantly, we filed out of the bathroom. No one uttered a word. It was time to get back to chemistry and algebra and geology.

All the stuff that didn't really matter, after you'd just realized your childhood dream.

☆ ☆ ☆

There comes a moment during a liftoff when you're forever seduced by the thrill and romance of space exploration. For WCBS radio news correspondent Christopher Glenn, the hook came when he covered one of the early Apollo flights. At the time, Glenn was a producer with WNEW radio in New York. While he and the other members of the press focused on the welfare of the astronauts and reliability of the rocket's hardware, the unexpected occurred.

Glenn and his news anchor were stationed three miles from the launch pad, sandwiched in a cramped trailer between the massive buildings manned by CBS, NBC, and ABC. Glenn was seated in the back of the room; the anchorman was in front of the trailer, before a large picture window. Glenn worked the dials and switches, ensuring the broadcast went smoothly.

The mighty Saturn lifted off in a tremendous explosion of flame, smoke, and sound. The anchor calmly recounted the launch, moment by moment, until, suddenly, everything went dead. As the rocket cleared the tower, Glenn realized that the trailer had lost its power. The whole operation had gone down in an instant, and the radio broadcast had inexplicably been terminated. Glenn rose from his console. He sneaked a peek around the room, which was now buzzing with excitement

What the hell's going on here? We're off the air!

Glenn searched the trailer. He had to restore power quickly, or the station would lose the entire launch broadcast.

He panned the room, his mind racing, pulse quickening. The rocket continued to climb slowly, shrinking with every passing second.

There! The outlet! Glenn thought. *The master AC plug has been pulled from the wall!*

Glenn sprang from his desk and raced over to the wall, where the main power plug hung. He slammed the plug back into the outlet, and power was restored to the trailer. The launch was nearly over, but Glenn and the anchor were back in business. The incredible vibration during the first few moments of the launch had shaken the trailer so violently, the power connector had popped from the socket.

It was then that Chris Glenn truly understood the power, the magnificence, of a live launch. Later in his career, Glenn would anchor many Shuttle launches and missions for WCBS radio. Each would be a source of inspiration; some, sources of deep concern.

"The Shuttle is very dramatic, very fiery, very loud and very quick," Glenn said. "It leaps off the pad. Whereas the Saturn . . . just sitting there and ponderously, slowly, inch by inch lifting off the pad, and this gigantic enormous sound just pounding you all over your body and in your brain. It's really dramatic and astonishingly emotional. You just sort of stand there, going *Get up, Get up, Get out of here!* At that point, when I saw my first launch, it just made an impression on me; I was forever hooked after that."

Liftoffs can also be cathartic for astronauts and reporters. On October 3, 1962, Wally Schirra became the fifth American to fly in space, the third to orbit the Earth. The moment *Sigma 7* cleared the tower, Schirra breathed a sigh of relief. This was *his* flight now: He no longer had to worry that a NASA administrator would bump him in the flight rotation, or that the flight surgeon would ground him for the most benign ailment.

"The one thing I felt was a sort of exhilaration, the euphoria," Schirra said. "Right after the liftoff, you knew it was yours, and nobody could take away that ride."

For 37 years, that sense of eternity has linked astronaut with reporter, with housewife, engineer, student. That first live launch shapes the way you see the world and yourself.

"I'll go out there whether it's 4:00 in the morning or 4:00 in the afternoon," said Howard Benedict, Executive Director of the Astronaut Scholarship Program and a former Associated Press reporter who covered dozens of launches. "I still get goosebumps when they lift off. I still think back to the Saturn V. When it lifts off, it is the most magnificent sight and sound in the world, I guess."

For CBS News Correspondent Scott Pelley, a Shuttle launch is the sunny afternoon following a devastating tornado. Once the network's chief space reporter, and now its top political correspondent, Pelley has witnessed the brilliance and potential of mankind in the raw power and beauty of a liftoff; he has also observed man's most desperate and depraved side while covering the Oklahoma City bombing trial and the World Trade Center bombing story.

"A Space Shuttle launch is a beautiful thing; it's something you can be inspired by, something that gives one hope," Pelley said. "I have the same kind of feeling when I see a physician save the life of a child, or when I see a tremendous piece of architecture. It is, in an important way, a reaffirmation of the good things that can be achieved by man in contrast to all the dreadful things one has to cover. It's uplifting."

To Pelley, launch day at the Cape is a spring Sunday morning in Florence, Italy. The technological splendor of the Shuttle is the magnificence of the Santa Maria del Fiore cathedral. The sleek, delta-winged orbiter is the stunning, gothic Duomo Santa Maria, the church's dome, inspired by architect Filippo Brunelleschi and constructed between 1294 and 1436.

"The cathedral was the unachievable, spectacular feat of its time," Pelley said. "It really gives one a sense of what achievement was, when you place it in comparison with the Space Shuttle of today. That's what I mean by the sense of knowing that man can do great things; it's when you see the Shuttle fly."

The great things include a Coast Guard sailor braving a raging sea to pluck a stranded seaman from a capsized boat; an emergency room physician saving the life of a child; counselors freeing a 16-year-old of a drug addiction. The heroes. The astronauts. They infuse Pelley with a sense of equanimity.

"Knowing that when I get through the next story on some dreadful event in human history, there will be another Shuttle launch, there will be another American hero and something to give me balance in my life," Pelley reflected.

Pelley, who is married and has two children, likens a Shuttle launch to the birth of a child. Both offer a sense of hope for the future, he said.

"When your children are born, you initially hope they're healthy and their mother is healthy. You begin a whole series of hopes upon hopes of what their lives will be like and what you want from them. In a sense there's a similar sense of progression with a Space Shuttle liftoff. You hope it's successful and that everyone survives and that the mission goes well. Then you start thinking what humanity can do with the space program and jumping from one hope to the other. A hope that man can reach out to each other and a hope that man can reach out to the universe uniting us here on Earth."

☆ ☆ ☆

If Scott Pelley's conclusion was sound, then ultimately Project Mercury was not only a boon to technology and space exploration, it was a gift to humanity.

Was STS-1 comparable in magnitude to the birth of a child? *Columbia* did unite thousands of strangers for a single purpose. The flights of the Mercury astronauts brought a nation together.

How was a birth similar to a rocket liftoff? I really wasn't sure. One does not do a great deal of musing during the interminable hours of labor and delivery. One does, however, ponder to the point of extreme introspection during the long countdown to a liftoff.

Ultimately, there was only one way to prove or debunk Pelley's judgment.

My opportunity came October 9, 1997.

☆ ☆ ☆

Take a deep, cleansing breath, I told myself. I should have given that advice to my wife, who was now breathless. She was the one about to have a baby, but you'd think it was I who was in need of coaching.

Supporting her with my right hand, we shoved our way through a set of heavy, tinted glass doors that ushered us into the main lobby of the Children's Hospital at Yale New Haven Hospital. Directly ahead was a large overhead sign that read: ADMITTING. An arrow pointed to the right. We inched our way through the central room, and I was flooded with a wave of relief.

Thank God we made it.

I feared that if we didn't get Debbie into the labor room quickly, I might have to scrub. Since Debbie's fingernails were buried a few inches into my neck, it was obvious the contractions were growing stronger. I led her quickly to the admissions office, which was little more than a large circular desk off one of the many hallways.

This would be our third child, and I was anything but accustomed to the torturous process of labor and delivery. It was like waiting for *Friendship 7* to survive re-entry.

In what seemed like the time it takes seven billion children to be born, an admissions representative took Debbie's name and other such trivial information, then turned and disappeared into an adjoining office. A few minutes later, she returned, pushing a folded wheelchair.

Debbie was screaming and crying on my shoulder now, so I frantically and spastically put my limited engineering background to work and pried open the wheelchair. I eased Debbie into the chair.

As we had done with Brooke, we chose to learn the gender of our third child prior to his birth. We would call him Adam, not in deference to the Biblical first man, but because Adam was a far better name than Trevor, our second choice.

As I held Debbie's hand, I recalled Scott Pelley comparing the launch of a rocket to the birth of a child. It was too soon to say if Pelley was just waxing philosophical and maudlin.

"Now hold on, honey," the attendant interjected. "I don't have time to stop and rub your back."

Suddenly, we were off and racing down the hall toward a bank of elevators. When we reached them, I closed my eyes and took a ride back in time to an era when 5'6" French-Canadian guys like me wore geeky cranberry-colored blazers, gray slacks and loafers, and manned rickety elevators eight hours a day.

Second Floor: women's lingerie, camping equipment, men's prostate gland stimulators. Third floor: sporting goods, floral arrangements, aspirin. Fourth floor: Newborn Special Care . . . Vaginal deliveries, Cesarean sections, post-partum, blood-curdling screams, 18 hours of pushing, placentas and umbilical cords for sale.

DING!

We emerged into a bright hallway; golden sunlight poured through a huge window on the west wall. I nervously checked my watch once again: 12:30 p.m. I felt like John Glenn stepping from the gantry elevator, capsule perched before him like some ramshackle kids' treehouse jammed into the crook of a giant oak. Glenn was lucky: he knew not what would befall him. I did: Labor and Delivery.

A moment later, we were in Room 466, which was all peach and cheery and bathed in invigorating white sunlight. The room actually smelled pleasant; it lacked the sour odor of waxy floors, the bitter smell of cleaning fluids. A single bed, moveable to any place within this private room, was stationed against the near wall. A bank of electronic fetal and cardiac monitors was recessed unobtrusively in the wall above the bed. Incongruously, a 19-inch television set was suspended from the ceiling above the farthest wall: Maybe Debbie would give birth during *One Life to Live.*

A few minutes later, a nurse breezed into the room. "Hi, I'm Judy. I will be your nurse."

Judy was about 45, attractive in a maternal way. She handed Debbie a hospital gown. "Put this on, Deb, and then I'll check you out." I shook my head; it seemed pointless for her to wear any clothing at all. But like a good patient, she slipped into the backless dress. Then I helped her into bed.

"How are you doing?" I asked Debbie, as Judy took her blood pressure.

"The back pressure is awful. I could take anything but that."

I knew it was time to put on my coach's hat. Soon, Debbie would be in transition labor; that's comparable to a Mercury splashdown, minus the parachute.

Debbie was now curled into a fetal position, as the contractions continued and sometimes coupled. Nurse Judy, meanwhile, patiently attached an external fetal monitor to Debbie's abdomen.

"I'm going to have someone take a look at you," Judy said. Then she left the room.

Moments later, a resident doctor, perhaps 30 years old, returned with our nurse. He examined Debbie meticulously. *How many centimeters is she dilated, Doc?* I looked into his eyes, which were doubtful, ponderous. *Is he solving some esoteric math problem? Does he know what he's doing?*

Finally, he completed his examination.

Give me seven centimeters, gimme seven, I thought.

"She's about six and a half," he reported to Judy.

Six and a half. Not bad. I felt my stomach quiver, my chin twitch. Tears were beginning to pool in my eyes. I fought them back.

There was hope. The same hope Scott Pelley felt while watching the Shuttle blast off before his eyes.

"Doctor Chelouche will be here in a few minutes," said Judy. "We'll have her check you again. You're really moving along!"

With that news, Debbie brightened up considerably. Birth *countups,* like launch *countdowns*, render one so helpless until the end of the preliminary stages. When the clock strikes zero, that's when an astronaut becomes a pilot. When the cervix is dilated 10 centimeters, that's when the expectant mother can go to work bringing her child into the world.

☆ ☆ ☆

At 1:05 Judy returned with Debbie's obstetrician.

"Hi, I'm Dr. Chelouche," she announced.

She was about 32, with liquid dark eyes, and locks of jet black hair that tumbled to her shoulders. Her skin was olive, her features dark. She wore a simple black dress that fell over a body devoid of curves and bumps. But it was her smile that arrested me: easy, confident, happy.

Dr. Chelouche slipped on a pair of sterile gloves and skated over to Debbie. "Hi, Debbie, I'm just going to take a look at you myself. Seems this baby really wants to come out."

Debbie smiled like a sick child who feels comfort in the presence of a person whom she trusts and loves. She scooted over, dragging the cords and leads of the external fetal monitor Judy had attached to her abdomen. That prompted the *beep-beep* sound of the baby's heartbeat to bounce crazily on the meter. Judy readjusted the wires and once again Adam's heartbeat came across nice and even.

I studied Dr. Chelouche's face, trying to read her thoughts.

"Oh, yes. She's seven and coming along nicely," she said

Seven!!!! Oh, Joy. We were at T-Plus Three in our countup to the birth of Adam. Maybe this would be easy, after all.

"Debbie, did you want some medication for the pain?" Dr. Chelouche asked.

"I was hoping for an epidural; I'm all ready when you are," said Debbie.

Dr. Chelouche smiled. "Well, we can try for the epidural, but I'm not sure we will have the time. I want to break your water in a few minutes; that should speed things up. Let me change and we'll take another look."

The closer we get to the end, the more the end runs away from us, I thought.

In the time Dr. Chelouche was gone, Debbie began to introvert. The more I wanted to help, the more she pushed me away. "Don't touch me. Just go away."

This is the time when you wonder if God really does exist. If he does, then why does he allow a person to suffer more than she can bear? "God never gives us more than we can deal with," some sages say. Of course, these are the same people who have never given birth.

At length, Dr. Chelouche returned, clad in gray hospital scrubs.

"I can't take any more of this back pain," Debbie cried. "Anything but the back pain."

I backed against the wall, feeling utterly impotent. Judy, sensing my helplessness, moved over and took Debbie by the hand. Words were useless at this point.

"I'm going to break your water now," advised Dr. Chelouche.

Judy helped Debbie lean back in the bed. Next, Dr. Chelouche pulled out what resembled a plastic knitting needle, about a foot long, with a small hook at the end.

As I stared at the tool that would puncture the baby's amniotic sack, my knees grew weak. I looked around the room; nothing seemed real.

The doctor slowly, patiently inserted the instrument. A moment later a stream of ginger-colored water gushed out and onto a few towels that had been placed on the bed. I noticed that the fluid was also tinted light green.

"There's some meconium in the fluid," Dr. Chelouche said calmly.

Debbie turned away, tears in her eyes.

"Oh, no!"

I stood there, dumbstruck.

"What's the problem?" I asked.

"There was meconium in the amniotic fluid," explained Judy. "That means the baby took a poop."

The news nearly floored me. Now I understood. Debbie looked up at me with desperation in her eyes.

Meconium is the first feces of a newborn infant; it is of a tar-like consistency, and if released in the amniotic fluid, it can choke the baby. I had heard of instances where babies choked to death on their meconium.

Debbie's womb, which had provided all the sustenance and protection our child required for 40 weeks, was failing him, much the way Americans perceived the heat shield had compromised John Glenn's safety within his steel womb.

Were we about to lose Adam? This boy, who already had two sisters who kissed him good night through their mommy's tummy: Would he never see his sisters?

"The color of the fluid is very light," added Dr. Chelouche. "So I'm not concerned with it."

So why was I still worried?

Just then, Dr. Chelouche sent my mercurial emotions on another roller coaster ride. "You're *eight* now," she told Debbie, smiling. "At this rate you should be able to start pushing in ten to twenty minutes."

She was wrong.

A half hour later, Debbie was still dilated eight centimeters. That prompted the doctor to order the epidural.

I closed my eyes and waited for the anesthesiologist to arrive. *God, let it be Steve,* I prayed. Steve was Dr. Steve Eige, husband of Dr. Nancy Brown, my children's primary pediatrician. Steve was a friend, Steve was a space buff, like I was. He had grown up flying model rockets, had once smashed his piggy bank to come up with the necessary funds to purchase the biggest and best model rocket around. "If you couldn't get on a rocket, at least you could build one as beautiful," Steve once told me.

He had grown up as a skinny kid who wore glasses and feared that, because of these disadvantages, he would never become an astronaut. He was right, but for different reasons. Somewhere along the line, he decided to become a doctor; now he was preserving life at this prestigious hospital and doing advanced research on medical something-or-others that were way over my head. Despite the seriousness of his work, he still lit up like a 10-year-old whenever he talked about astronauts and rockets and the Moon.

Suddenly, I was roused from my thoughts by the clanking of shoes across the tiled hallway floor. As the clackety-clack grew louder, I opened my eyes, searched out the door, and saw a man bustle into the room. *It's not Steve,* I lamented. The man was tall, cocoa skinned, and had probing dark eyes. He introduced himself, but because of his thick accent (and my remaining mental fog), I was unable to distinguish his name. It sounded something like Dr. Mert. *Okay, Dr. Mert, or whatever your name is, do your thing, and do it quickly.*

☆ ☆ ☆

"I can't take the pain anymore!" She was curled over a table, her hands clutching Judy's. Her knuckles were white and her fingertips were bright red from squeezing so hard.

Behind her, Dr. Mert probed her spine with his fingertips, searching for the proper spot to insert the epidural needle. But he was a desperate prospector, panning for gold that did not exist.

To exacerbate my dismay, Judy announced that she was leaving; her kids were waiting for her in front of their school. A new nurse entered the room, and Judy introduced us. Her name was Beth, a bubbly woman who looked to be in her late 20s. She took Debbie's hands in hers, while I stroked Debbie's head.

A moment later, another anesthesiologist joined our expanding group. He introduced himself as Dr. Sinatra, and then consulted with Dr. Mert. A look of concern spread across his humorless face.

And that's when I spied the epidural catheter sitting in a metal tray, on the table before Debbie. It was about 18 inches long, and I could not believe anyone could possibly insert it into a woman's back.

"Jesus Christ," I muttered.

"Are you okay?" Beth asked. She rubbed my left arm like a mom comforting her frightened child.

I could only stare blankly and wonder when this nightmare would end.

Dr. Sinatra waited for an opportunity to insert the needle into Debbie's back. It was impossible to complete the delicate procedure with Debbie twisting and bending during a contraction. He worked patiently and with great care. Slowly probing, checking. Moving the needle around her back.

"Tell me where you feel the needle," he told Debbie.

"I feel it on the right side. Oh, I can't take it anymore! Hurry!"

Dr. Sinatra paused during the contractions then raced the clock and Debbie's pain to find just the right spot to insert the catheter.

It's never going to get done. They'll never get this damn epidural working and Debbie's going to be in big trouble.

"I'm moving it to the left a little, Debbie. Tell me where you feel the tingling," said Dr. Sinatra.

"It's still on the right side," she cried.

Come on, pal. We can't take much more. Make it happen this time.

Another incapacitating contraction. Beth and I held Debbie's hands as her back arched through another contraction.

"Where do you feel it now?"

"It's equal."

Once more, Dr. Sinatra inserted the needle. He paused, threaded the catheter through the needle. He slipped the needle back out of Debbie's spine. Then the doctor taped the catheter to Debbie's back to prevent it from being pulled out. Now the sedative could flow from a pump, through the catheter, and into Debbie's bloodstream. Shortly, she would feel little, if any, pain.

The worst was behind us.

Or so I thought.

As the medication began to take effect, Adam's and Debbie's vital signs began to drop.

I perked my ear toward the fetal heart monitor. *Beep-Beep-Beep-Beep-Beep-Beep.*

The echo of the baby's heartbeat.

120 beats per minute.

A moment later: *Beep—-Beep—-Beep—-Beep—-Beep.*

90 beats per minute.

"Hang in there, honey," pleaded Debbie.

I watched in horror as Adam's heartbeat slipped to 80 beats per minute. I knew normal was 120 to 160. Adam was in trouble.

Worse, Debbie's blood pressure was dropping; it was now 100/65.

"Let's give her some oxygen," Dr. Sinatra ordered.

The nurse slipped an oxygen mask over Debbie's face.

Her blood pressure: 90/60.

Beep——Beep——-Beep———Beep———-Beep————Beep.

Adam's heart rate: 70 beats per minute, and sinking.

I'm going to lose them both. I've come here for the miracle of birth; instead I am going to be forever changed by an unexpected death. It's like watching Challenger *explode before my eyes.*

Dr. Sinatra remained calm. He filled a syringe with a clear fluid, then he injected it into the intravenous pump.

"It's important we keep the baby's heart rate up," he explained.

I listened for the staccato beeps.

Seconds passed. No change. A few more seconds . . .

Beep—-Beep—-Beep—-Beep.

Seventy-five beats per minute.

C'mon, Adam, keep it going.

Beep—Beep—Beep—Beep—Beep.

Eighty beats. Then 90, then 100. Then, finally, smooth and steady.

I peeked at the monitor above Debbie's head. Her blood pressure: 120/70.

"Perfect!" exclaimed Dr. Sinatra.

I allowed myself a deep breath. I was soaked in nervous perspiration.

Behind the transparent oxygen mask, Debbie's mouth was curled in a smile.

☆ ☆ ☆

A few minutes later, Debbie was resting comfortably. The doctors had left, but had advised us they were within shouting distance.

As luck would have it, the revolving door of nurses continued to turn. Beth, our second nurse of the day, yielded to Chris, a stern-looking brunette.

"How are you feeling, Debbie?" she asked.

"Well, I'm having some pressure in my bottom, but at least the epidural has taken away the back pain."

"Okay; sounds like the baby is making its way down," advised Chris. "You should be able to start pushing soon. I'll have the doctor check you again."

Dr. Chelouche, who had been in and out for the last hour or so, returned. She donned her gloves and began to check Debbie.

I held my breath.

She said, "You're fully dilated and fully effaced. But the baby's still a little bit high. Why don't we give it a little while before you start pushing. That will allow the baby to drop and it will make it easier for you to push."

Ten. T-plus Zero. The countup was done; now Adam was John Glenn in a holding pattern aboard Friendship 7.

☆ ☆ ☆

"How's it going?" I asked Debbie, some time later.

"My right side burns and I have a lot of pressure in my bottom. But at least my back isn't killing me."

"Sometimes there's a window where the medication doesn't reach," explained Chris. "If you roll over on that side, it may help."

Debbie did as advised, but it didn't help. In fact, as time passed, the burning in her right side grew into a dull ache. I noticed that her eyes were no longer sharp. She seemed to be slipping back into a fog of introversion.

"How are you now?" I asked a few minutes later.

"My right side is really starting to ache and my bottom has a lot of pressure."

"That's not right. The epidural should make you completely numb."

"Don't you think I know that?" Debbie snapped.

"I'll go get Chris and Dr. Chelouche," I said, and raced off down the hall.

Chris had left a few minutes earlier to review Debbie's medical history. Dr. Chelouche was jockeying a few patients. It was one busy day on Labor and Delivery.

I found Chris at the nurse's station.

"Debbie's getting a lot of pressure and she says her right side is really aching."

"OK. I'll be right there."

When I got back to the room, Debbie had resumed her breathing regimen. I took her hands in mine.

"Don't touch me. Please . . . just leave me alone right now. My right side is killing me."

Chris entered the room and began to massage Debbie's trembling thighs.

"Your right side hurts, Deb?"

"Yes, it's starting to feel as bad as before the epidural. I don't think the epidural is working anymore."

"Try lying on your right side," advised Chris. "Maybe you have to push? I'll get Dr. Sinatra and see if he wants to give you more medication."

Good idea, I thought.

Minutes later, Doctors Sinatra and Chelouche returned. "Where are you feeling the pain?" asked Dr. Sinatra.

"In my right side, all the way down my leg. The pain is awful." Once again, Debbie's face was as white as a sheet and was twisted in a mask of agony.

Dr. Sinatra adjusted the IV pump to increase the flow of the medication.

"Give it a contraction or two to take effect," Dr. Sinatra advised.

Six contractions passed, with Debbie screaming under the full power of the labor pains.

"I can't go through this back labor again!" Debbie cried through coupled contractions.

"The only thing I can think is that the catheter somehow slipped out," said Dr. Sinatra. "Then, Deb, you know you have two choices: We can do the epidural again, or you can go without it and grit it out through labor."

Some choices, Doc.

We had to make a decision quickly, for Debbie could not bear this pain much longer. For a few minutes, we discussed the options. Neither was optimum. Finally, Chris suggested a possible solution.

"Deb, why don't you try to push a few times? Maybe the baby has dropped low enough for you to push him right out."

"Good idea," I agreed, as did Dr. Sinatra, as he left the room.

Chris raised the back of Debbie's bed; Dr. Chelouche placed Debbie's feet in stirrups.

"OK, Deb, take a deep breath and push," encouraged Dr. Chelouche.

I grabbed Deb's right leg from behind the knee and pulled it back toward her head. Chris did the same with the other leg. Debbie, now a pro at pushing, breathed deeply and pushed from the tops of her eyebrows. She shoved with all her might, her face violet with strain.

"Good," said a smiling Dr. Chelouche.

"Again," I commanded.

Another huge push and then another. Debbie's bottom was extending, swelling and spreading wider and wider now. It was doing a thousand things you'd never repeat at the dinner table.

Dr. Chelouche checked her.

"OK, Deb, hang in there."

Suddenly, there was a flurry of activity. Bright lights illuminated the whole room. It felt as if we were onstage at some surreal 19th century medical freak show.

Dr. Chelouche leaped into a blue surgical gown and slipped her feet into protective shoe booties.

Launch time.

I was infused with an energy not of my own source. Adam was coming, and on some deeper level I could sense him telling me so.

Come on big guy. We're ready for you. We all want to see what you look like.

Dr. Chelouche slipped a sterile mask over her nose and mouth and then what resembled a plastic welder's mask over her eyes.

"OK, Deb. Start pushing again."

Chris and I pushed Debbie's legs against her chest. Deb gave a mighty push. Then another, directly behind it.

"Really good job," Chris and I said in unison. Dr. Chelouche wore a smile of anticipation.

Another huge push.

"I can see his head! He has blond hair!"

"You're almost there, Deb," urged Chris. "Another big push and maybe the head will come out."

Debbie shoved and pushed and slowly she opened up all the way. A head, or, rather, a squished, slimy, blood covered sponge-like organ was slowly emerging.

"His head is out!" I cried. "Keep pushing!"

While Debbie pushed, Dr. Chelouche twisted and turned Adam's head, working to ease the rest of his body out, ever so slowly. This was a *Lift Out*. But the doctor was struggling, I noticed. Adam was one big boy, and while his head popped out easily, his shoulders appeared to be stuck.

"I have a shoulder here," barked Dr. Chelouche. The next voice I heard was Chris calling down to the nurse's station: "We have a shoulder here."

Instantly, a man with golden hair and a ruddy face, dressed in nurse's garb, stormed into the room. Without uttering a word, he leaped onto Debbie's belly. As Debbie pushed, and Dr. Chelouche pulled, this man pressed down firmly on Debbie's abdomen. He was literally trying to shove Adam's shoulders out!

"I've got it!" said Dr. Chelouche.

First she wiggled out one shoulder, then the next. Finally, with a tremendous pull, she wrenched Adam free.

"Debbie, Adam's here! Hi, Adam!" I cried.

I did not see angels, nor did I hear trumpets blare the moment our son entered our world. But I did feel a curious closeness to God. In the moment before two pediatricians took Adam to be suctioned, I understood how Father Tom Egan claimed John Glenn had experienced the knowledge of God by flying through the heavens. With the birth of Adam, I felt I was experiencing the grace of a higher source, as well.

Then I remembered the pediatricians. We still had the issue of the meconium with which to deal. As Debbie cried with joy, I heard the wail of my son for the first time. I looked at the clock: 4:40.

A few minutes later, I held in my arms my nine pound, 10 ounce son. He was all pink cheeks poking out the top of a blue blanket. A few yellow curls peeked from under the brim of a tiny white cap. He looked just like his sisters: pug nose, pointed chin, a sardonic smirk on his face. Kris Kringle, as a newborn.

I kissed Adam on the forehead, then handed him to Debbie, who was now being stitched and cleaned by Chris and Dr. Chelouche.

Then I stepped back and drank in the whole scene.

As I watched these people, all sunny and bright, grinning like kids at my son, I was struck by a revelation: This wasn't just the birth of a child; this was a celebration, a communion of people all pulling together for a single purpose.

Adam's birth, like the flights of *Friendship 7* and *Columbia,* gave us that singular sense of hope.

A few minutes later, Dr. Sinatra returned, bearing the same somber expression. He glanced over at Adam, who was sleeping contentedly in Debbie's arms.

For the first time today, Dr. Sinatra smiled.

"He's a beautiful boy."

Then he left the room, never to see us again.

Five hours earlier, I had known none of these people; now I felt a kinship with them.

For every Dr. Sinatra there was a Chris Glenn; for every Dr. Chelouche, a Scott Pelley. A person for whom each childbirth or each launch *was* unique, with its own set of singular circumstances.

Scott Pelley was right: A launch and a childbirth made friends of strangers, made time disappear, made God a member of the team.

Funny, even the most defensive, argumentative, neurotic fathers-to-be warm up at *T-Plus 10.* Even the staunchest, most staid reporters and NASA employees experience that same timelessness and sense of hope when a countdown reaches Zero.

As we prepared to move upstairs to a private room, I pondered what life would be like if every person, each and every day, could experience the profound joy of a birth or rocket liftoff.

I patted Adam's head. *Maybe you'll grow up to be an astronaut; maybe the first man to step on Mars,* I thought.

I was waxing ingenuous now, seeing life through the eyes of a little boy.

Suite: Judy Blue Eyes

The six-day voyage of STS 41-D was significant on a number of fronts. First, it marked the inaugural flight of the Shuttle Discovery, *NASA's third orbiter. Moreover, the mission included the second American woman to fly in space, 35-year-old Judy Resnik. Resnik garnered newspaper headlines when she deployed the largest structure ever built in space, the 102-foot-long prototype of the Shuttle Power Extension Package solar array. The array was designed to provide additional electricity to the spacecraft, which would in turn permit longer missions.*

The mission, which ran from August 30 to September 5, 1984, was deemed by NASA one of its most successful Shuttle flights ever. But it was not without incident. During the flight, a 20-pound, 18-inch long icicle formed on the orbiter's exterior, causing NASA and the astronauts some consternation. They feared that if the icicle snapped off during reentry, it could damage the exterior of the craft and some of its protective tiles. With the advice of a ground crew that included Sally Ride, the first American woman to fly in space, commander Henry Hartsfield used the Remote Manipulator System (robot arm) to dislodge the block.

☆ ☆ ☆

"Dear God, please deliver me from this," singer/songwriter Jeannie Cunningham cried. Her head was numb, her body and soul wasting away from a nine-year cocaine addiction that had culminated in this three-day binge to oblivion.

She toasted her 28th birthday alone, in her Los Angeles home that was perilously close to entering bank foreclosure. A bottle of tequila and a Valium served as presents, celebration of a life skidding utterly out of control. Cunningham stumbled into her kitchen, saw her reflection in a window—92 pounds. So gaunt, so devoid of life, so wasted away. *Just like my life.*

Where had it all gone? she wondered. *It went up my nose,* she responded grudgingly to the judge and jury in her conscience. She was unable to do anything about that fact of life, however. Her house had all but gone up her nose, along with the money she had saved to build a recording studio. She had even pawned her dead mother's jewelry to raise money, but had squandered that away, also.

Sad truth of the matter was, she'd snorted cocaine so frequently, she had literally blown out her nose. Her septum was tattered, mangled like that of a mediocre boxer. Her respiratory system was hardly that of a prizefighter's, either. She had gone into respiratory failure a year earlier, but doctors resuscitated her and sent her on her way. "You're as good as new," they told her. And two weeks later, she was shoving more white powder into a nose nearly devoid of nasal passages. Tonight, she battled a staph infection around her ears and hairline; yet, somehow, an occluded part of Jeannie Cunningham's will clung to life.

Please, God . . . come on!

Her heart was racing, a wild drumbeat in her chest. She swallowed the Valium and drew deeply on the bottle of tequila, thinking, hoping, the combination of depressant drugs would ease her racing heart, quiet her tumbling, rambling mind. Then she switched on the television and sat down. Her world was a blur now, and the voice of the newsperson on the screen droned into her throbbing head.

Shuttle . . . Discovery . . .

She caught bits and pieces of a report that stole her breath. Then, suddenly, an image popped onto the screen. Cunningham's sleep-deprived, sunken eyes opened wide.

Maiden voyage of Discovery, the newscaster spoke on. *Judy Resnik . . . second American woman to fly in space . . .*

The television camera zoomed in on a smiling woman with a striking shock of brown hair; she was floating weightless in the Shuttle, and she looked so peaceful, so at ease. *She grinned like a child without a care in the world. Judy Resnik . . .* Cunningham mused. *A woman in space.*

Cunningham had always been fascinated with space exploration, even in the days when a single astronaut pried himself into a capsule the size of a TV dinner. But that was a time when macho test pilots ruled space, a time when Jeannie Cunningham was just discovering her world.

Fighting against the cocaine-induced fog that was suffocating her mind, an awestruck Cunningham marveled at Resnik. The astronaut was deeply tanned, clad in blue astronaut shorts and polo shirt; she wore aviator glasses and looked to be all of 30. It was as if she had taken a giant leap off a California beach and landed upon a star.

Cunningham struggled to focus on the fuzzy television picture. *Waste management system . . . icicle . . . damage . . . Shuttle. RMS system to dislodge it . . . Dr. Judith Resnik . . .*

Suddenly, Cunningham was struck with a revelation: "Oh, my God, that's who I should be!" But no one was listening. "Look at this; this woman's doing it! Look how I've wasted my life!"

It was so cool! Resnik was just one of the guys—*but what a smile!* —the same as any other astronaut who had dared to visit outer space. It was just like her own life, Cunningham thought; on the road, playing music with five guys. All locked in together. Resnik was handling it so easily, just as Cunningham once had, when playing music was a joy, not torture on her vocal cords... when writing songs was a conduit to a greater purpose.

Finally, the broadcast concluded. But the image of Judy Resnik stayed in Jeannie Cunningham's mind. *If she can do that, I can kick cocaine.*

Soon, she fell into a restless, drug-induced sleep.

☆ ☆ ☆

The drugs had followed the pain, but it took Jeannie Cunningham many years to recognize why she ached inside.

Adopted at birth by a military family, Cunningham bounced from town to town, following the career of her adoptive father, a Marine officer. The Cunninghams barely had enough time to unpack their bags before Lor Cunningham was reassigned to yet another nondescript Marine base in another nondescript town. Or worse, in the bullet-ridden jungles of Vietnam.

Making friends was difficult for Jeannie Cunningham, but not only because her address changed with the seasons. She had been born with a club foot and walked with a pronounced limp until she was 16, when surgery corrected the defect. While kids shook their booties in go-go boots, fishnet stockings, and mini skirts, Cunningham's father forced her to wear inexpensive, plain dresses, lacking character or individuality.

Her father's military discipline and rigidness transcended mere style. Lor Cunningham was not an indulgent man, particularly to Jeannie and her brother Caleb, who was also adopted. Theirs was a "Yes, sir; No, sir" world, and if they disobeyed, they got spanked. As they grew older, their transgressions were met with a stern grounding.

Lor Cunningham demanded from his adopted children the same dutiful obedience he expected from his soldiers. A disciplined person was a responsible person, he reasoned. A child who followed the letter of the law grew into a self-sufficient adult. Lor Cunningham wanted the best for his children, even if they did not always understand his motives.

Sometimes these lessons of tough love were difficult for Jeannie Cunningham to accept. As a child, she was afraid of the dark and often pleaded with her father to sleep with the lights on, but her father, and mother, Joan, refused her.

Not one to desire an imaginary friend, Jeannie sought solace from the brightest and most comforting light she could imagine: that cast by God. This beacon would be her guiding light in a life that would be frequently plunged into the darkness of desperation.

Jeannie Cunningham adored Lor Cunningham, despite her father's tough demeanor. This was, after all, the same man with whom she had prayed by her bed since she was two.

She had learned about God at Sunday school; *I can't see God, but maybe he can follow me,* she hypothesized.

One night, dad gave his daughter a gift to which she would often turn, even in her most helpless moments.

"God's got you tucked in his shirt pocket," Lor Cunningham whispered sweetly. "You're safe."

Thereafter, whenever she went to bed, Jeannie always pulled her covers up under her chin.

I'm safe now ...

If the military's maxim was *Look inside yourself for strength,* Lor Cunningham's advice to his daughter was the opposite: *Look toward the stars, to the heavens, for your answers.*

He imparted this bit of personal wisdom not through words, but through lessons.

One such message came when she was four. On a mild, dreamy evening, Lor Cunningham lay on the roof of his house with Jeannie and Caleb by his side.

"The first person to spot a star out there gets a quarter," he offered.

Jeannie's eyes went wide as she peered into the blue sky, growing pale, soon to blacken into darkness.

Come out come out wherever you are!

The three star-gazers remained as still as soldiers hiding in enemy territory: The stern commander, his silent, respectful children. Time passed, night slowly fingered its way toward them.

Finally . . .

"I see it!" cried Jeannie. Up in the sky, the night's first star winked at the Cunninghams.

Dad smiled. "Because you were the first one to find the star, you get to make a wish."

Jeannie closed her eyes. Lor Cunningham sang: "*Star light, star bright, first star I see tonight. I wish I may, I wish I might, have the wish I wish tonight.*"

Jeannie Cunningham became a quarter richer that night. She also made a pact with herself to always gaze up at the stars and make a wish.

Twinkle, twinkle, little star, how I wonder what you are?

Not only did her dad introduce Jeannie to the stars, he acquainted her with another celestial neighbor. One other night, he took his daughter outside to admire a new moon.

"Look Jeannie, God clipped his thumbnail," said Lor Cunningham quietly.

The little girl peered into the sky. The sun had just set, and, sure enough, the Moon did look like God's thumbnail!

"Will it fall?" she asked her dad, all wide-eyed innocent. "I don't want it to fall on me!"

"No, it's in space," reassured her dad.

And from that moment forth, Jeannie Cunningham was always astonished that nothing ever fell from the heavens.

As she grew up, she also marveled that people could put a man inside one of those big rockets and shoot him at that same giant thumbnail. And get him safely back home!

She was six the day Alan Shepard turned most Americans into ogling children. Even the normally staid Lor Cunningham was amazed with the feat. "I can't believe it!" he gushed, as he walked into the house that historic May 5, 1961. Nine months later, John Glenn circled the Earth three times, which really sent her dad for a loop. Moreover, if dad was excited, Jeannie was excited. After the flight, she went outdoors and gazed up into the starlit sky. *Who was that falling star?* she mused. *It is so amazing that this guy could go up to the stars!*

Slowly, Jeannie Cunningham began to make the connection between the sky flyers and the stars toward which they soared.

Two years later, Lor Cunningham was stationed in Vietnam. Back in San Francisco, his wife lay in a hospital, dying of mononucleosis, hepatitis, and pneumonia. Alcohol had ravaged her body, and its clutchhold was reaching toward Jeannie's throat as well. The way Jeannie Cunningham saw it, dad was the "baby killer," as some antiwar protesters called the American troops; he wasn't coming home. Mom was dying, she wasn't coming home. Soon, she feared, she and her brother would be orphans.

Jeannie sought refuge in the *stars*. But San Francisco was so smoggy, so riddled with pollution, she could seldom spot them. On sullen evenings, when she was sure she would break down and sob, Jeannie went outside and stared toward the heavens.

Dad, you're so far away. Can you see the same stars I see now? If I make a wish, will you hear it? she wondered, unconcerned about the difference in time between California and Vietnam. But that didn't matter. The stars would listen; God would hear.

Mom, please live. Dad, please come home.

☆ ☆ ☆

Lor Cunningham did return home, to wage a bedside vigil with his dying wife. Somehow, the wishes of a little girl were heard: Joan Cunningham survived, and, in time, recovered. With his wife healthy, and a semblance of familial normalcy returned, Lor Cunningham had a choice to make: accept a second military tour stateside and be with his family, or return to Asia.

Feeling he could better serve his country overseas, Lor Cunningham packed his bags and returned to Vietnam.

It was a decision that would undermine the stability of his family.

He retired from the military in 1967, returned home, grew a beard. As if serving penance for being away from home, he became a social worker. But while Lor Cunningham was able to help society's ills, he was powerless to reverse the downward spiral into which Joan's life had spun. She was drinking heavily now and was irrational most of the time, even a bully toward her daughter.

As Jeannie learned one night, merely leaving a shirttail poking from the top of the clothes hamper brought heavy repercussions: Joan Cunningham glowered with sadistic satisfaction as her trembling 11-year-old daughter emptied the entire hamper at 2 a.m. one night, then replaced all the dirty clothes. Stultified, Lor Cunningham stood by, said nothing.

By 1968, Jeannie Cunningham could no longer bear to live at home. Her mother's abuse, coupled with her father's failure to defend her, had made life intolerable.

There was only one refuge: boarding school.

How many 12-year-olds beg to be sent to a strange school, miles from home? During five of the next six years, Jeannie Cunningham would bounce from boarding school to boarding school, continuing the transient military life. When money was short, however, she returned home and spent the 1971 scholastic year at a public school.

She had friends now, but the wrong friends. Kids who consumed drugs as if they were candy. Soon, Jeannie began to follow their lead. At age 15, she could do a whole rack of "White Crosses"; that is, in one day, she was consuming 11 pills of the amphetamine better known as speed.

Her grades, always a strength, began to decline. "How could you do so well at boarding school, but struggle at a public school with a weaker educational system?" Lor Cunningham asked.

Jeannie had always considered her father a man who appreciated honesty. Now she told the truth.

"Dad, I've been doing drugs. I can't handle it. I don't feel like I belong on the whole planet."

The look on her father's face was a sickening mixture of disbelief and horror. Jeannie had let him down. Now, Lor Cunningham had a 15-year-old daughter addicted to drugs and a 42-year-old wife addicted to alcohol and dying of cirrhosis of the liver.

"Can you get off this stuff?" he asked his daughter.

She didn't know.

Her father took a week off from his job in child protective services. He kept her off the phone, away from her friends, in her room. Fed her soup. There was no temptation; but at night, when it was dark, Jeannie had that internal *light.*

Somehow, at the end of the week, Jeannie Cunningham was drug-free. As a last act of volition, Joan Cunningham vowed to send her daughter back to boarding school.

That's when Jeannie Cunningham's grandmother came to the rescue. Ruth W. Akin, "Nonna," had not always agreed with her granddaughter's decisions, but she always provided unconditional love. She had given Jeannie her first Bible and her first guitar. To Jeannie, Nonna was God personified, incapable of passing judgment on her fellow man. Able only to love.

Now, she would be her daughter's guardian angel by paying for her private school education.

☆ ☆ ☆

Jeannie Cunningham was 16 the day Joan Cunningham died: February 17, 1973.

Jeannie had been singing and playing in talent shows since age 12, and now, she figured, was a good time to put her talents to good use. When asked to write a song for her mother's upcoming memorial service, Jeannie Cunningham eagerly agreed.

But what should she write? She pondered a handful of ideas, rejected them all, grew frustrated. She slammed her guitar against her bedroom wall.

While she struggled to find her voice, her brother Caleb and sister-in-law Cindy were en route to her home in Twentynine Palms, California.

The last 24 hours had been melancholy for them, as well. A night earlier, while driving down the highway, Cindy had asked Caleb to pull over to the side of the road.

"Look on the horizon," she said. "There's a star twinkling that reminds me of your mother's eyes."

Caleb Cunningham shared little of his sister's fancy or spirituality. Still, he humored his wife and looked into the California sky at the star that blinked at him. He continued to stare at this unique star; he was transfixed by its hypnotic beauty. *There's something calming about it,* he thought.

They arrived at his father's home a night later, flying into the driveway, the car's tires screeching loudly.

Cindy and Caleb Cunningham jumped out of the car, then raced inside to get Jeannie.

"Mom's outside!" Cindy cried, while dragging Jeannie outside.

Oh, My God, Jeannie Cunningham thought. *Mom's dead. I'm going to see a ghost!*

"Look on the horizon," Cindy ordered, pointing at the sky. "Is there anything you notice that's different? The one star blinking, that one. Does it remind you of anyone?"

Cunningham stared into the sky. Then she felt a chill run through her.

"Yeah, it reminds me of mom's eyes when she was healthy!" she said.

She raced inside the house. A minute later she was back with her father, now red-eyed from crying.

"Dad, look up in the sky. Do you see anything weird?"

This was the man with whom she had once sat atop their home, tossing wishes toward the stars. Would he spot that one star out of so many that blinked at them?

"Well, I notice one star that reminds me of your mom's eyes," he responded finally. A small smile stole its way onto his face; his eyes became warm, soft. Jeannie Cunningham looked lovingly into those crestfallen eyes. *The star had comforted him.*

A few minutes later, Lor Cunningham led his children back into the house. He drew an astronomy atlas from the bookcase.

"Let me see if we can plot it for future reference," he decided. Cunningham determined the star's location, then pushed a pin into the map, at the constellation *Aquarius*.

Baffled, Caleb and Jeannie Cunningham turned and faced their father; they looked into his surprised eyes. Aquarius had been Joan Cunningham's astrological sign.

Before falling asleep that night, Jeannie Cunningham found the inspiration for her mom's musical tribute. She would call it *Twilight.*

☆ ☆ ☆

Lor Cunningham's period of mourning lasted less than six months. Shortly after burying his wife, he met Elizabeth Fitzgerald on the beach at Cabo San Lucas. The couple fell in love and were married a few months later.

It didn't take Jeannie Cunningham long to conclude that her stepmother eschewed children; *Betty doesn't wish to share her new husband, even with his children*. As in the past, in times of parent/child conflict, Lor Cunningham stood silently by his wife's side.

Just as quickly, Jeannie began to dislike Betty Cunningham. Boarding school was a refuge; when she was not there, she stayed with Caleb or Nonna. Finally, at Nonna's behest, Jeannie

Cunningham transferred to an American school in Lugano, Switzerland. It was, in some respects, a sanctuary, removed from familial discord, away from her own mental and emotional static. Just school, music, friends, a new world. She held concerts for her friends, she rallied the kids against the teachers. She recorded war songs, even a score for a musical version of *A Midsummer Night's Dream*. At the conclusion of the year, she won a music award and graduated with honors.

Her graduation gift came from Nonna: a round trip ticket for Lor Cunningham to visit his daughter in Switzerland.

During her year overseas, Cunningham experienced yet another life-changing incident, when she attended a rock concert by Ike and Tina Turner. When the stately and impressive Ike Turner took the stage, Cunningham thought he was *coooool*. But *Tina* Turner; oh those legs, and that energy and fire!

This rocks my world! she thought. *Somehow I have to meet her.*

Jeannie Cunningham also sensed that Tina Turner would play a role in her becoming a musician. *If* she could just meet the rock diva. Cunningham considered her options and her contacts around Europe. A few days after the concert she learned, through parents of friends, that in a few weeks the Turners would be playing in Zurich.

To raise the money for the concert tickets and train fare from Lugano to Zurich, Cunningham wrote term papers for her classmates. Money problem solved, she then appealed to the school for permission to go. "Enjoy yourself," said the headmaster.

Next hurdle to climb: getting backstage to meet Tina Turner. But Cunningham had a plan, albeit a partially-baked plan: *I'll get to Zurich, buy some flowers, and write a poem for her.*

Two weeks later, Jeannie Cunningham was alone, on a train bound for Zurich, determined to meet Tina Turner. During the trip, she struck up a conversation with another passenger, a young woman. The woman knew the location of the concert hall. "I'll escort you to the show," she offered to Cunningham, who was most happy to accept.

The pair arrived at their stop at 6:20 p.m., nearly three hours before the start of the show. Cunningham panned the immediate area and was disappointed to see that all the flower shops were closed. *My plan is falling apart,* she lamented. *What do I do now?* Her resourceful friend had the solution.

She led Cunningham back through the train station to a small flower shop that, like the others, was closed. But that didn't daunt her friend; the determined woman pounded on the glass window, and a moment later, the door opened. Minutes later, Cunningham emerged from the shop with a gorgeous bouquet of roses.

Cunningham's travel companion led her to a trolley that would take them to the concert hall. Minutes later, the trolley car came to a grinding halt. "This is where you get off," her friend advised. Cunningham thanked her and jumped off the car. She walked around to the front of the trolley car, stopped, turned around. But the woman was nowhere to be found. Cunningham searched the train, the entire area: no one. *This guardian angel has taken me all the way and helped me find flowers when all the shops were closed,* she thought. *She led me to the place, and now she's gone.*

Cunningham shrugged, then checked her watch: 7:30. Two hours to go. She entered the concert hall and found it deserted, save for the band's road crew. Mustering all her courage, she approached one of the roadies.

"Excuse me, how can I get these flowers to Tina?"

"I'll take them," the young man responded, and reached for the flowers.

"Hold on a minute, let me write a card," interjected Cunningham. She quickly scribbled a note: *Roses are red, violets are blue, no other fan could be as devoted to you.*

Cunningham explained that she was an American schooling in Switzerland, and that Tina's music was a source of inspiration. The star-struck girl finished the card with: *So that you'll recognize me, when you play* I Get By With a Little Help From My Friends, *I'll come up and present you with one red rose.*

Cunningham saved one rose for that song; the others she gave to the roadie to take backstage to Turner. Then she took her seat, far back in the building, and waited . . .

☆ ☆ ☆

Three chords into *I Get By* . . . Jeannie Cunningham was shoving her way down the long aisle and racing toward the stage. Amidst the pounding roar of applause, she muscled her way to the stage. But there was one problem: The stage was about six feet high and Cunningham was a hair over five-feet-four. "Tina, Tina!" screamed Cunningham amidst the din. She frantically waved the single rose above the plane of the stage. She continued to shout Turner's name; she continued to wave the rose at the singer.

Please see it.

A burly bodyguard gave her a half-hearted shove, but Cunningham refused to budge. Suddenly, just when Cunningham was about to give up, she caught a glimpse of Tina Turner's stunning legs strutting her way.

Then the rose and card were disappearing into the singer's slender fingers.

"Are you Jeannie?" Turner asked as she bent over the edge of the stage.

"Yes," Cunningham responded meekly.

"Thank you for the flowers!"

Tina Turner reciprocated by dedicating the next song to her number one fan, Jeannie Cunningham.

With the assistance of another obliging stage hand, Cunningham met the dynamic rock stars after the show.

"You're a really good writer," said Ike Turner, referring to Cunningham's poem. "You look us up when you get back to California. We have a studio in L.A. Give us a call." He handed Cunningham his business card.

Jeannie Cunningham was about to enter the elite circle of Ike and Tina Turner. Little did she know, however, that within that circle waited the demon that would take her life on a downward spiral.

☆ ☆ ☆

After graduating from high school, Jeannie Cunningham returned to California seeking a new direction in her life. She lived with Caleb for a few months, but found she needed more. She wanted to be a musician. And she knew who could help her accomplish her goal: Tina Turner. She recalled Ike Turner's invitation that memorable night in Zurich: "*When you get back to California, look us up.*"

I'm going to Los Angeles to become a rock star.

Cunningham packed a week's worth of clothes and the $750 she had earned from writing papers for her classmates. She grabbed her Les Paul electric guitar and a 12-string Ovation acoustic guitar. She stuffed Ike Turner's business card into her suitcase. With Brett, her Great Dane/Afghan mix, as company, she sped off in search of stardom, sputtering forward in a beat up, army green Chevy van, and headed to the home of Beach Boys keyboard player Carly Munoz.

Cunningham had been introduced to Munoz by Susan Ford, daughter of President Gerald Ford. The two had become fast friends when the Cunninghams were stationed in Virginia, and Munoz had invited Cunningham to stay with him in L.A. if she was in town. When she arrived at Munoz's house, she discovered things were more complicated than she had envisioned. Munoz never told Cunningham that he had a girlfriend; Cunningham never admitted to Munoz that she owned a dog. The Munoz welcome mat was being stepped upon by too many feet . . . and paws. Cunningham thought: *I was raised never to impose upon people.* The solution was simple enough. Cunningham and Brett turned their van into a mobile home and moved into Munoz's driveway.

I'll be inventive, Cunningham decided. After all, she had been a Marine Corps child and was used to compromising and improvising. She had a nice, comfortable bed in the van, and access to Munoz's bathroom and telephone. Best yet, Munoz had a high-tech recording studio in his home, the best Cunningham had ever seen. So things weren't so bad.

Yet.

For five months, Cunningham and Brett lived in this microworld on the outskirts of city blocks wherein criminals ran rampant. Cunningham avoided the street gangs, but she could not ignore the seedier elements of the music industry. She secured a job as a gym instructor; she drove a cab, joined a band. Saved her money so that one day she could live in a place where the bedroom and bathroom were under the same roof.

She contacted the Turners at their Los Angeles studio, Bolic Sound. "Come on over," they responded. When Cunningham arrived at the studio, Ike and Tina Turner were recording a song, and Tina was belting out the lyrics with her trademark youthful energy. Ike Turner smiled at Cunningham and motioned for her to join him. He extended his hand to her, and in the palm was an oddly shaped object, with a straw extended from it.

Cunningham, wishing to fit in, acquiesced; she inserted the end of the straw into her nostril and breathed in deeply.

Moments later, the cocaine took effect.

My eyeballs are going to pop out!

That day marked the beginning of the end for Jeannie Cunningham.

Nevertheless, she grew to adore Tina Turner who, Cunningham noted with irony, never touched drugs. Cunningham covered the van's walls with posters of the famous rock singer. Cunningham began hanging around with studio sycophants who were along strictly for the drug ride. The crowd soon became antisocial, downright weird.

The more people Cunningham met through Ike and Tina, the more phone calls she received in the middle of the night. Those late-night intrusions angered Munoz's girlfriend to no end.

Cunningham figured it was time to move on. She rented an apartment and soon met a young man named David. The couple fell in love, moved in together, and got engaged, all within a few months' time. It was now June, 1976; the wedding was scheduled for August.

But before Jeannie Cunningham would ever wear a bridal gown, she learned another painful lesson about life.

One day, while at the recording studio, she struck up a conversation with one of the guitarists in the Turner band. Cunningham complained that her Les Paul was unwieldy on stage; he had an Ovation that she liked. "Why not swap guitars?" he suggested.

They celebrated the exchange over dinner. At the end of the evening, the guitarist invited Cunningham back to his home.

Great, I'm in the inner circle, Cunningham thought.

Suddenly, when they were inside the guitarist's home, he pulled out a butcher's knife. The enraged man stabbed wildly at Cunningham's shirt and pants, then began shredding Cunningham's clothes. But in the midst of this assault, he lost his balance; the knife cut through Cunningham's side, between her ribs.

The terrified young woman struggled to break free.

"Try to scream and you'll never scream again," her assailant threatened.

He thrust the knife against her neck. Jeannie Cunningham was utterly defenseless now . . .

☆ ☆ ☆

The rape devastated Cunningham physically and mentally. Worse still, she lacked the confidence to press charges: *It's always the woman's fault,* she thought. Cunningham recognized that she had put herself in a tenuous situation by going to her assailant's house. By virtue of being in his home, she believed she had granted permission for the rape to occur.

Like cocaine, the assault left her numb, and no longer able to respond to her fiancé's touch. The couple canceled the wedding, one month before its scheduled date, and they broke up shortly afterwards.

By the time Tina Turner left her husband, Jeannie Cunningham was a full-blown cocaine addict.

Nearly a year passed, with Cunningham striving to do something with her fledgling music career. In June, 1977, she formed an all-girl band with three other musicians, called the "Hollywood Stars." They took their act on the road and were signed by a Japanese promotions company. A tour of Japan sounded like a good idea to the band members, at least at the outset. But, upon arrival in Japan, the promotions team took the musicians' passports, essentially rendering Jeannie and her crew prisoners, unable to leave the country.

The promoters demanded that Cunningham change the name of the band to '"The Cherries," which, in Japanese, means *nipples.*

"You will play in clubs and you will be quiet," they ordered. "Otherwise, we will have the Mafia take care of you."

The band was powerless now, forced to play music in clubs of the promoters' choice. The women were also expected to be hostesses, who allowed patrons to fondle them at will. After her rape, the thought of a stranger touching her made Cunningham physically ill.

On the final night of their scheduled tour, the band was playing on a stage some six feet high. They were in the midst of a song when a burly man, clearly drunk, staggered past two Japanese bodyguards. He pulled himself onto the stage; he grabbed Linda, the keyboardist.

"I want to fuck you," he shouted in broken English. Then he reached out and grabbed her breast. Linda, still playing the keyboards with one hand, whacked his hand away.

Despite the interruption, the band continued to play, barely missing a beat. Still, the inebriated man stumbled forward, now toward Suzy, the drummer.

"Hey, someone get this guy outa here!" Cunningham yelled to the bodyguards. But they were bent over, howling with laughter.

The drunk tripped over Suzy's drum kit, sending a pair of cymbals flying.

The band played on.

Suzy, following her musical cue, performed a perfectly synchronized drum roll. But when she got to the spot where the cymbals normally were, she instead found the man's exposed knuckles. Like Buddy Rich flying around his kit in a furious blur, Suzy raised her sticks and slammed them down on the man's hand as hard as she could. The man howled in pain, grabbed his injured hand with his good hand. The band played on. The audience screamed with laughter.

The man stumbled over to the bass player, Oneida. He dropped to one knee in front of the microphone stand. He moved his head forward, sticking his face as close as he could to the guitarist's crotch. Oneida recoiled.

The band kept playing, but now they were short a singer and unable to execute their trademark three-part harmonies.

Cunningham continued to sing, to strum the guitar with her left hand. With her free right hand, she pushed the microphone stand back to Oneida. The man crawled over to Cunningham. With his uninjured hand, he reached up and squeezed Cunningham's breast with all his might.

The pain was excruciating, the humiliation overwhelming.

Despite her physical and emotional agony, Cunningham reached back, and with a powerful backhand, sent the man flying across the stage.

She laid down her guitar. The band members stormed off the stage.

The crowd cheered, laughed, shouted for more.

The band members had nowhere to turn. Later that night the owner of the café apologized to them for the unruly man's lewd behavior; but the band soon learned that he had also apologized to the man for Cunningham's aggression. Worse yet, the proprietor gave the man a "bottle-keep," a whisky bottle, mounted on a plaque, with the man's name engraved on it.

Over the next few weeks, Linda would all but go off the deep end. The day following the incident of the drunken patron, she purchased a bottle of laundry detergent. The club in which they had experienced the ignominious incident boasted an impressive water fountain outside its main entrance. During the day, when the club was closed and no one was around, Linda poured the entire bottle of soap down the fountain's main spigot.

Hours later, when the fountain was activated, the club transformed into a mountain of suds, a bubble bath for the entire city. The women watched with mirthful delight as the club's managers scowled, and people slipped and flopped around the rooms. No drinks were poured that night.

Two weeks later, Cunningham sent Linda home. She, Suzy, and Oneida would also escape from Japan, but not before their Japanese promoters tried to set them up on a drug bust.

When Jeannie Cunningham returned to the United States, she had lost her faith in humanity. She was 21, a drug addict, her mind tortured by a series of real-life nightmares. Caleb and Cindy had two kids now. Nonna was getting old. Worse yet, Bolic Sound had burned down. Her former sanctuary was gone.

If this is the best I can see in music, I have nothing to live for.

Yet, she trudged on. With the money she had earned from the Japan tour, Cunningham bought a house; she continued to sing, and also worked odd jobs. But years of drinking, smoking cigarettes and marijuana, and doing cocaine was taking its toll on her health and her career. For the better part of the next seven years, Cunningham played in bands, performing Top 40 songs in half-empty hotel lounges. The band often played five sets a night, and by the end of the evening, the singing, the drugs, the booze had rendered her nose and throat so raw, it pained her to even consider breathing through her mouth.

Despite her physical and mental anguish, Cunningham continued to sing, continued to *exist. If my own mother died from drinking, what's the difference?* Cunningham told herself. The truth was, she could not make it through the long days, the interminable five-set nights, without a lift. Jeannie Cunningham's life became a battle between light and darkness; her appetite for cocaine had become uncontrollable. Her career was going nowhere. And yet, that *light* inside of her, the one that always connected her to the heavens, urged her to keep trying. That mental compass to the stars still pointed north.

Deep down inside, Cunningham knew she would never commit suicide, would never end up naked on a hill, lost in a drug fog.

But God's light was shrinking into a tiny flame inside of her.

There were nights when the numbness wore off, when the vial of cocaine was empty and Jeannie Cunningham reached to God. When it was 2 a.m. and she was so wired she was in that limbo between being too far awake to sleep, but still having enough energy to think *If I had some more coke, maybe I could chuck out another song.*

On those occasions, the coke was wasted, and pleas to God were useless. So Cunningham did the next best thing: She grabbed a bottle of tequila and took a Valium. Then she sat and waited for the drugs to take effect, waited for her world to become *real* once again. She breathed through a parched mouth, and her nose felt like there were two cement trucks parked inside. She blew forcefully, saw blood on the tissue; her head throbbed, and she sat in a stupor and thought: *This is one of the best times to find God. Why am I doing this?*

Those nights, she eventually fell asleep, but not before seeking God somewhere among the stars. *Please, God, help me. God deliver me.*

After nearly nine years of vials and bottles, joints and pills, Jeannie Cunningham finally was delivered. The night of her 28th birthday.

☆ ☆ ☆

Jeannie Cunningham desperately wanted to *feel,* but her personal demon would no longer allow her to do so.

Some eight hours after falling into a Valium/cocaine/tequila-drenched sleep, Cunningham awoke, still feeling foggy. Her first instinct was to pick up the phone. She started to dial the number of her drug dealer, but her fingers wouldn't work She couldn't even remember her supplier's number. That had never happened before. She pulled herself out of bed, stumbled into the kitchen, and made herself a cup of coffee.

The caffeine began to revive her and the demon that hungered for her soul. *I have to get him quick,* she thought frantically. But she had no cash, no way to pay her dealer for even the smallest amount of dope. Did she have something else to sell, something that could buy her a fix to get her through another day? She panned her house, searching for valuables.

As she searched, her dealer's telephone number popped into her head. She grabbed the phone, started to dial. Suddenly, she froze.

Right there, in front of her eyes, was a mental vision of astronaut Judy Resnik, smiling, floating around in space as cool as Cunningham was when she hung out with her fellow musicians before a gig.

I don't want to live like this, Cunningham screamed inside her head. She wanted to be a Judy Resnik, someone who had contributed to the world and had made a difference in other people's lives. Someone who saved the day, as had Resnik during the flight of *Discovery.*

She put the phone back in its cradle. *If I have to do it later, I'll call later.*

She got by on will for the next few hours, but her tenuous courage began to wane. Once again, she decided to contact her drug dealer. Once again, the image of Judy Resnik stormed into her mind's eye.

The next few days were grueling for Jeannie Cunningham; but she got by and turned her back to the phone. Instead of thinking in terms of survival, she began to think in terms of goals.

After a few weeks of cleaner living, she returned to another addiction, music. Twenty years after composing her first tune and strumming it on a homemade cardboard-and-rubber band guitar, Jeannie Cunningham was on the road and touring with Lionel Richie. Shortly after, she met up with David Crosby, of Crosby, Stills, Nash, and Young, whom she had met through Houston-based disk jockey Dayna Steele. She went on the road with Crosby. She never looked back. Whenever the demon whispered in her ear, the image of Judy Resnik chased it away like a lice-infested street bum.

She never forgot the significance of STS 41-D. *I'm one single woman in the United States who saw a broadcast of another woman in the United States doing heroic things with her life. Goddammit, why am I wasting my life? If she can do it, so can I.*

Music once again became Cunningham's life blood; space exploration, her passion. What was better than sitting around a room with a bunch of musicians pondering life as an astronaut? As the weeks and months passed, Cunningham had ample time to conduct impromptu meetings with her Mercury Generation buddies.

She also had enough money and work to enable her to crawl out of debt. She saved her home and started making plans to build her own recording studio. And she pledged that one day she would meet Judy Resnik, to thank her for saving her life. But Cunningham knew she would not do so until she was successful, on even terms with the astronaut.

Two years later, one of Cunningham's dreams blew up with the *Challenger.* Seven astronauts died that day, including Christa McAuliffe, the first teacher in space, and Judy Resnik. Resnik's death devastated Cunningham; it robbed her of the opportunity to meet the astronaut and to present her with a song she had written in her honor. Even as the wreckage of the spacecraft spilled into the Atlantic Ocean, Cunningham's telephone rang: "Are you okay?" her friends asked. Cunningham knew what they meant: *Are you going to revert to doing cocaine?*

Her close friends recognized that when Cunningham was sad, which was not very often, she became self-destructive. One of the things that had enticed Cunningham to break her cocaine addiction was the anticipation of meeting Judy Resnik. Cunningham had put pictures of Resnik on the walls of her studio and she had bragged to her friends, "Now that girl knows how to live; she works with guys like I work with the guys. She plays piano, she sunbathes. She is a normal girl. She is someone who likes science, someone who likes the stars!"

Now, the person whom Jeannie Cunningham most admired was gone.

Because Cunningham still drank beer and wine, and smoked marijuana occasionally, her friends wondered how difficult it would be for her to reintroduce a hard drug such as cocaine.

Her inspiration's gone, what's she going to do now? they thought.

Instead of falling backward into a snow drift, Cunningham marched outward and onward, past her old, destructive habits. *I will move forward, I certainly will not go back now,* she vowed. *And I will keep Judy's memory alive.*

She took a major step forward by naming her new recording studio Resnik One. Next, to Resnik's family, she forwarded a copy of the musical tribute. In no time, she had befriended the Resniks.

☆ ☆ ☆

Fate intervened once again, when Cunningham met another distinguished Shuttle astronaut, Bonnie Dunbar, veteran of the 1985 flight STS 61-A, the West German D-1 Spacelab mission. During the flight, Dunbar was responsible for operating Spacelab and carrying out a variety of experiments.

If Resnik was Cunningham's model and sentinel, Dunbar was her surrogate big sister. To the admiring Cunningham, Dunbar was another heroic woman, another astronaut who flew on the coattails of Shepard and Glenn and the rest of the Mercury 7 pilots who put their butts on the line. Dunbar's accomplishments encouraged Cunningham to put her soul on the line, to stay away from drugs, to continue to sing professionally, to record music.

During a career that would include three other flights, Bonnie Dunbar also gave Cunningham another gift, the joy of experiencing a live liftoff: The leap off the pad, first a low rumble, and then the explosion of sound and power and fury, bearing into her chest. That was the technological aspect of a liftoff. But Cunningham was also bidding *bon voyage* to a traveler who was also a friend, teacher, companion. First came the concern for her friend's welfare, anxiety over the outcome of the mission. Then came the tears, first from relief, then from the exquisite beauty of the machine soaring into the sky, confirming the incredible integrity of the technology. *All of these people working together, working with the* crème de la crème *of society, test pilots.*

Today, astronauts and space exploration are as much a part of Cunningham's life as music. NASA has become her family, the space pilots her role models, and to stumble would be to fail in their eyes. To maintain a sense of equity, she has given back to the space program by recording songs that have been taken on Shuttle missions.

"They're tremendous role models, and it saved my life," said Cunningham, now 42, and often on the road, where she is known to her fans as "Jae Cie." The influence of Dunbar and Resnik has expedited the healing process; three years of therapy has also played a big part. Most important is her faith in God. Cunningham can quote passages from the Book of Genesis as easily as she can recount the details of a Shuttle or Mercury mission. She reads voraciously books on spirituality, and can discuss the Dead Sea Scrolls and the teachings of the Buddha or Christ as comfortably as a theology student.

Through the dark years, there was always that tiny flicker of light inside her, saying *I am with you.* Many nights, she was that fragile child, pulling the covers up around her chin, waiting for God to clutch her in His pocket.

Today, her life's maxim reads: *Seek and ye shall find.* Or, as the Buddha would say: *Knock and the door shall open.*

Over the last few years, the door has been wide open for Cunningham. When not recording, touring with her band, or watching space missions, Cunningham writes "rah-rah" tunes for companies such as Chevron. It's a good life, she says. Ask her what she'd be doing were she not a musician, and she responds: "I'd be an astronaut."

Cocaine and its trappings are a thing of the past. So is marijuana, as of 1991. Now, she turns to other sources for comfort. Space exploration continues to serve as a source of inspiration. Through flight, Cunningham finds hope for the future.

She still flies, but no longer because of drugs. A licensed pilot, she flies a Piper Cherokee 180 and is licensed to fly a helicopter. To be able to soar among the clouds is another measure of her personal growth: just a few years ago, Cunningham was terrified of flying. She overcame her phobia because to live in fear, she decided, was to not live at all.

She has learned the power of creation rather than destruction. Disciplined diet, rigorous exercise, and a weightlifting regimen have instilled in her a resolve that what one builds up, one will not destroy.

Through the turmoil and the rehabilitation, she is, in many ways, still age four. Still prone to gaze up into a star-specked sky to seek out that one brilliantly blinking star. Maybe it will be the "Aquarius star," the one that twinkled like her mom's eyes. Maybe another; doesn't matter which, for as Christ said: "My Father's house has many mansions." To Cunningham, those "mansions" are the countless star constellations. All of those many, many stars still respond to her wishes, still flicker an acknowledgment whenever Jeannie Cunningham thinks, *When you wish upon a star . . .*

☆ ☆ ☆

It was the destiny of mankind.

At least that's what singer/songwriter David Crosby had concluded. Space. The great refuge of man, removed from the bunged-up mudball that was the planet Earth, a world he felt certain would one day self-destruct.

Intelligent man was drawn to space, Crosby had also decided. Certainly, Crosby had been drawn to the wonder of the heavens, other worlds, other possible life forms. He had long been attracted to space exploration. But whether David Crosby, the walrus-mustachioed, potbellied member of the legendary band Crosby, Stills, Nash, and Young, was intelligent depended on one's definition of the word.

On a day of great hope for mankind, and for the future of space exploration, David Crosby was recuperating in the infirmary of a Dallas jail. He was serving a five-year sentence (which would be commuted to 11 months) for possession of a Colt .45 handgun and for free-basing cocaine in a Dallas nightclub. The self-described "chemical dumpsite" huddled around a television with 30 other convicts. This was not the vantage point from which he would have chosen to watch an event as historic as the liftoff of the Space Shuttle *Challenger*: The flight marked the first time a civilian would fly in space, teacher Christa McAuliffe. Also on board was Judy Resnik, for whom Crosby had great admiration.

If incarceration had chained the bodies of Crosby and the other inmates to their cells, it had not entrapped their souls. A buzz of excitement and anticipation filled the room as the mighty *Challenger* rose off the launch pad and into the blue Florida sky.

Then, in one horrifying moment, the blue sky became spotted and streaked with white smoke and fragments of the spacecraft. In the blink of an eye, the lives of the crew members had been extinguished. Thirty-one men, each of whom had had their lives taken from them by a court of law, watched in stunned silence as seven astronauts had their lives seized from them by the failure of a seemingly infallible machine.

A little bit of David Crosby died on January 28, 1986, along with Judy Resnik and her colleagues. *These space pilots were the people who were able to dream and were great at inspiring a dream in the rest of the world,* mused Crosby. Suddenly, that dream had disintegrated.

He was a dreamer before he became a 20-year member of the 1960s. Space flight captured the fancy of a young David Crosby at an age when kids sneaked Superman comic books under their bedspreads and read by the light of a flashlight. By age 10, Crosby was reading the works of science fiction writer Robert Heinlein, as well as many of the pulp science fiction magazines that would transform impressionable kids his age into space voyagers.

His dreams held no images of pretty freckle-faced girls with pigtails and smiles that made you blush until your ears were red. No, slumber led him to worlds of different cultures, where man learned the many different forms life could assume. Where there were societies from which we could learn. Earth and Mars, neighbors.

Project Mercury seemed a natural evolution to Crosby. *This is what we should be doing.*

If many Americans saw Mercury as the solution to the Communist Red Scare, Crosby considered the program the remedy for man's dark side, the animal within, bent on destruction. Mercury was the antithesis of the "let's throw a bomb at somebody" mentality. It was the "let's see what's out there in the universe." If the threat of a Russian nuclear attack made your mom and dad look for answers within their hearts, Mercury encouraged people to look outward, to the future of mankind.

These were baby steps, a glorious beginning to a new era away from self-destruction.

He might have been one of them: a pioneer trekking to different worlds, or the ambassador of this "blue marble planet." But there were the turbulent '60s, and, for David Crosby, protracted adolescence.

"By the time they made the first flights, I was already in so much trouble, they never would have taken me," said Crosby with a laugh. "By the time I got out of trouble I was no longer good enough to where I could hope. But I just dream about space exploration for mankind, I dream about it for my kids, I dream about it for the future."

Space exploration has been a staple of his life, as has music and the hit records that long ago made pop icons out of him and his band. *Suite: Judy Blue Eyes; Woodstock; Carry On; Teach Your Children.* The spirit of a carefree generation.

Crosby was able to meld his two passions years ago, when Crosby, Stills, & Nash invited a group of astronauts, including Bonnie Dunbar, backstage during their concert in Houston.

"Bonnie Dunbar is a great role model for so many people in this country today," said Crosby. "She has all the brains and all the strength and all the character and still is this very feminine woman and real bright. She's everything I think is admirable for all women."

When not discussing missions with current astronauts, Crosby often finds himself lobbying in the political arena for additional NASA funding. "Every time a senator or a congressman comes to me and says 'Hey, I need your help in that direction down in . . . and we're really up against some terrible people,' I answer: 'What's your position on the NASA budget?' I think, 'Christ, we don't need another B2 bomber, we need to go to the Moon, we need to go to the fucking asteroid belt.'"

At age 56, Crosby is healthy and happy. A 1994 liver transplant gave him a second chance at life. Then there's his 2-year-old son Django.

"There's something to be said about survival, especially when you get kids," he said. "They make it all worthwhile."

A look into his son's eyes and Crosby sees tomorrow. Space exploration, he says, is the panacea for mental enslavement.

"From my earliest days my dream has always been that mankind would grow up and go out to all the galaxies," he said. "That we would stop wasting our time killing each other and go out and spread out in the universe. If you include the rest of the universe in your perspective, the petty bickering that goes on here starts to look pretty small and it makes you want the human race to grow up."

☆ ☆ ☆

T*his is incredible; it's extremely claustrophobic in here,* radio talk show host Dayna Justiz Steele thought. Around her, three NASA technicians pried, pulled, and shoe-horned her into the world's most cramped spacecraft, a NASA spacesuit. Although she was by no means a large woman, the extra large EVA suit still was a tight fit. A *stiff* fit, to be precise. *This is uncomfortable!*

It should have been. What she wore was a virtual environmental protection assembly, a 12-piece suit, including helmet and a liquid cooling and ventilation garment that contained some 300 feet of coolant tubes. It was her entire life support system.

My God, this is not fun and games when you're outside the Shuttle and into the darkness and vacuum of space, she thought.

Here, *Astronaut* Dayna Steele didn't have to worry that her helmet might float away in the weightless environment of the Shuttle.

Here, safe in a technician's room at Houston's Lyndon B. Johnson Space Center, she had ample support personnel eager to shove her unwieldy shoulders into her suit, like an obstetrician squeezing a baby's shoulders through his mother's birth canal.

Finally, with a twist, the huge bulky helmet was in place, and Steele suddenly was immersed into a world silent, save for the rhythm of her breathing, the thumping of her heart. It was like scuba diving, the sense of claustrophobia, the sense of isolation.

This is what it feels like to be an astronaut, to go out there . . .

☆ ☆ ☆

Dayna Steele had David Crosby to thank for her day in ersatz space. She had befriended the musician when Crosby was in a Houston area halfway house completing the remainder of his prison term. Upon his release in 1986, Crosby contacted Steele, and the two soon met, to talk. Eventually the subject switched to space exploration and NASA.

"You mean you have lived in the Houston area your whole life and you've never been to NASA?" Crosby asked, stunned.

Steele had grown up in Seabrook, Texas, some 26 miles from Houston. Although the town was home to many astronauts during their NASA days, Seabrook and astronauts were to Steele as Mickey Mouse and Disney World are to lifelong residents of Orlando, Florida: a fact of everyday life.

"Growing up around NASA, well, I always knew NASA was around here someplace," she said.

Such bold insouciance nonplused Crosby, who considered astronauts mini-gods.

As a child, Steele and her family had lived down the street from Apollo 7 astronaut Walt Cunningham and his family. These days, it was no big deal to see last week's Shuttle pilot walking his dog on a sunny Saturday morning. In fact, Steele had even confused her neighbor, veteran astronaut John Young, with the one homeless man in Seabrook.

Stunned, Crosby replied: "I'm going to arrange a tour of NASA for you."

In early 1987, Shuttle astronaut Bonnie Dunbar gave Steele more than a tour; she provided a hands-on review and inspection of the hardware, and an explanation of how it worked.

I can't imagine I lived here my whole life and have never been to NASA. I can't believe I went to Houston schools and never took a field trip to NASA, Steele thought, at the conclusion of her visit. *I went to work today and was on the air for a few hours. These people would go to work and blast off a rocket into space!*

Suddenly, Steele felt as if *she* had just flown in space. Blind to the exploits of the astronauts for most of her life, Steele now considered manned space exploration of vital national importance. *We just ignore this . . . This amazing group of people was so close to me and I had never known it my whole life, and we were actually up there!*

Crosby and Dunbar had opened Steele's eyes and heart to the magic in her own backyard, to the significance of liftoffs and landings and astronaut welcoming parties. It was so easy to go back in time, now, to her childhood, and the flight of Apollo XI. To see herself sitting anxiously before the family's television set, mom and dad, brother Scooter, Happy the dog, Fluffy the chicken, and Grandma Nanaw by her side. Then Steele was holding her breath as Neil Armstrong stepped stiffly, slowly, down the ladder of the lunar module.

"That's one small step for man . . ."

It all made so much sense to her now; it all seemed real. So real, in fact, that shortly after her visit to Johnson Space Center, Steele began to put her celebrity clout to good use. When musicians came to town, she set up NASA tours for them. The Devil's Advocates quickly became NASA cheerleaders when Steele explained that their treasured electric guitars and the computer-generated sound mixers would never have existed were it not for space research.

By the conclusion of 1990, astronauts and space flight were as much a part of Steele's life as stale cups of coffee and Camels were to an Apollo flight director. In June, Crosby returned to Houston with fellow band members Stephen Stills and Graham Nash.

"There are a bunch of astronauts who want to come to the show, and there's a bunch we want to see," Crosby told Steele. "Can you be the liaison?"

To the musicians, it was a given that while astronauts could perform a pinpoint landing on the Sea of Tranquility, they couldn't navigate their way backstage even with compasses strapped to their wrists. Steele gladly accepted her assignment and went to work arranging tickets and backstage passes for the astronauts. Included in the NASA entourage was her future husband, NASA research pilot Charlie Justiz.

☆ ☆ ☆

T*his will be a marriage made in the heavens*, Dayna Steele knew. *If he ever proposes to me, and I know he will.* She and Charlie Justiz had dated for months; now Steele was eager to get married. On July 4, 1991, she had discovered an engagement ring in Charlie's pants pocket. But days passed and no wedding proposal. The weeks dragged by like a prison sentence. Still no proposal. August rolled around and Dayna Steele was frustrated to tears. She decided to confront her boyfriend.

"I won't be badgered into marrying you," Charlie explained, tersely.

"Oh, yes, you will!" she retorted, regretting immediately her hastiness.

Then it dawned on her: *He has a plan . . .*

It all made sense now. Astronaut Tom Henricks, who was living with them at the time, was scheduled to fly in an upcoming mission.

I bet they have the ring packed away on the Shuttle, and when it's packed away, you can't get at it. She grinned with the satisfaction of a little girl who discovered her birthday gift hidden under her parents' bed. But soon that smile became a frown. Delays set back Henricks' mission four months. By December 1, 1991, the Shuttle astronauts had completed their mission. And so had her engagement ring . . . she hoped. Still, Charlie had shown no sign that a marriage proposal was imminent.

Finally, just when Steele was about to launch her fiancé into low Earth orbit, Charlie Justiz presented her with an early Christmas gift. Tears sprang in Steele's eyes when she read the inscription on the engagement ring: *A hundred and nine times around the world for you. Love, Cisco.*

They were married three months later.

☆ ☆ ☆

"I feel like I've been in outer space," Dayna Steele says today. Hardly a week passes when one of her neighbors or former housemates isn't squeezing his or her way into one of those Slinkyish 12-piece EVA suits. Steele treats each mission as if it were the flight of *Freedom 7* or *Friendship 7*. These space voyagers are, after all, her friends, her spiritual family. You can bet they will return home to a hero's welcome, if Steele can help it.

"On Friday, at four o'clock, the astronauts return home to Ellington Field," Steele will announce to her listening audience. "It's open to the public and anyone can go. It's absolutely free, and you can welcome the astronauts home."

Some folks heed her heartfelt notice and attend. But, mainly, the welcoming committee consists of the friends and families of the astronauts and crew. There used to be a high school pep band at these celebrations, but no more. But there are lots of American flags to be waved, and many more flags line the streets where the astronauts live. Some residents, such as Steele, recognize the significance of the just-concluded flight. Others stare at the flags and wonder aloud: "It's not Flag Day, why are there flags out?"

This is not to say interest in space flight has died. Cooled, perhaps. Five years ago, Dayna Steele coveted a NASA patch as if it were a Moon rock. Today, such memorabilia are used as cold drink coasters.

Then there is the story of "Dack."

He was born Charles William Justiz, but on this side of the Moon, no one refers to him by his birth name, not since that memorable NASA party at astronaut Ed Lu's house.

Dayna Steele was five months pregnant with her second child, and she had grown tired of referring to the baby as "It," or "Pubert," unflattering nicknames by any stretch of the imagination. During Lu's party, one of the NASA surgeons approached Steele. He patted her on her belly, good-naturedly. "The baby needs a good NASA name," he declared, then paused to consider a suitable name.

"We're going to call him *DACK,*" he concluded, finally.

"Dack?" Steele wondered.

"Yeah, that's his unofficial NASA acronym: *Dayna And Charlie's Kid.*"

Charles William Justiz was born seven weeks premature and weighed just four pounds, seven ounces. Some six hours after his birth at Columbia Clear Lake Medical Center, mom and dad made their way down to the intensive care unit, where their tiny baby was being monitored.

Dayna Steele took one look at her son's incubator and broke into peals of laughter. An identification card posted on the side of the baby's bed read: Charles William "Dack" Justiz. "Well, I guess his name is Dack," Steele said, grinning.

☆ ☆ ☆

Two years later, Dack Justiz bears all the earmarks of a future NASA flyer. His NASA patch collection is the envy of every two-year-old in Seabrook, and he shares his dad's fascination with aircraft. Maybe he'll grow up to be a NASA research pilot like his dad, teaching a new generation of flyers how to land the most advanced spacecraft. His mom smiles hopefully at the thought.

"I hope he carries on the dream," she said. "I really do."

☆ ☆ ☆

The funniest calls came from the new mothers.

"Hey, Dayna, can you get me a space suit for my son?" they'd say. "I know you have connections with NASA. Maybe some sort of astronaut pajamas. Something cute."

Fours years ago, she and her husband could only respond, "We'll see what we can do."

Today, they chuckle aloud and answer, "Would you like a long-sleeve terry flight suit pajama?"

At the Justiz's Internet outlet, The Space Store, boys and girls of every size, shape, and age can dress and feel like Gordon Cooper, John Glenn, or Sally Ride. Hats, T-shirts, sweatshirts, ersatz flight food, mission contraband of every kind are distributed from here. In fact, for a lucky few space travelers, Steele will even bring the merchandise home and personally pack it. Especially if the recipient is an 18-pound four-month-old who curls his upper lip at the sight of strained peas.

But you can't visit the Space Store unless you own a computer and have access to the Internet. Then, a quick dial up to the web site www.thespacestore.com and the universe is yours.

☆ ☆ ☆

On a Sunday afternoon a dozen days before Christmas, 1997, Dayna Steele and Charlie Justiz were enjoying a rare day of rest at their home in astronaut land. Justiz and Dack were in the living room, reveling in the latest installment of the syndicated science fiction television program *Babylon 5.* Steele, meanwhile, was awaiting a visit from a few astronauts who planned to exhaust the balances on their credit cards by ordering Shuttle Approach and Landing Simulator programs, NASA drink insulators, and black NASA golf hats.

Steele giggled guiltily; soon she would fill another box with a blue junior NASA flight suit, complete with flight patches and a requisite air of nobility. *I get such a kick out of sending people these little baby outfits. Friends who have just had a baby are fascinated.*

Even Scrooge's bachelor brother Grump would smirk seeing a pudgy, 20-week-old, red-haired, blue-eyed boy stuffed into an "I Want to Be an Astronaut" T-shirt. Boogers running down the kid's nose, spitup spewing out a pink-O mouth, globbing up and drooling down the embossed insignia of a Shuttle astronaut floating beyond the craft and the planet.

It was great, actually, watching astronaut moms and dads, so stoic when flying around the Earth in a spaceship, reduced to silly five-year-olds giggling red-faced over flatulence, as their kids tried on infant short-sleeved footless flight suit pajamas with the American flag on the left shoulder. That was the beauty of space exploration. It kept these people forever young.

☆ ☆ ☆

For Steele and Justiz, the creation of the Space Store was an extension of their own passion for space exploration. From a business perspective, the store came about thanks to the inevitable laws of supply and demand. For years, Justiz and Steele had fielded requests for NASA merchandise, but had no legitimate means to supply the product themselves. In December, 1997, they decided the market was right for them to go into business. A few days after opening the web site, the first orders began to filter in.

"The future of our country is kids," explained Justiz. "They should know the options. Half of the store is designed to sell products, the other half is to answer questions."

Justiz hopes that, in time, a 12-year-old would-be pilot will be able to access the web site and pose technical questions to specialists. He says "rocket scientists" will be used to answer such questions as: "How does the solar system work?" or "What are the chances of a comet striking the Earth?"

That the Space Store gears its line of products and services primarily to children and young adults is no surprise. Through his many public appearances at schools, Justiz has learned that students have a lot of questions and a lot of doubts about pursuing a career at NASA.

Justiz had many of those same questions when he was a child, following the Mercury and Gemini projects.

☆ ☆ ☆

The day *Sputnik I* sparked nationwide Red Paranoia, five-year-old Charlie Justiz came home from school to find his parents fearing for the future of their country. Although they lived in Miami, Florida, Charles and Gloria Justiz were native Cubans who frequently returned to a homeland caught in the throes of sweeping political change.

"This is terrible," lamented Charles Justiz. "It's the worst thing in the world for the Russians to beat us into space."

It was a *calamity.*

Even as a youth, Charlie Justiz sensed his parents' passion for space exploration, particularly when John Glenn and Alan Shepard helped mitigate the effects of the *calamity.* His father, a surgeon, called the flight of *Freedom 7* man's first step toward conquering all of life's great barriers.

We've conquered space, now cancer's next, thought Charles Justiz.

Space flight helped open a new world for Charlie Justiz, one teeming with model rockets and airplanes. The model airplanes were great: a fly-by-wire type that sent you spinning around and around in circles until you threw up and sent the plane spiraling wildly into a backyard grave. One day, this junior pilot flight training would serve him well.

For Charles Justiz, Shepard's mission was a way off the great rock called Earth. We weren't monkeys anymore, and that one 15-minute flight had separated us from all other animal life on the planet. We had proven ourselves to ourselves, and to a force greater than us.

Tomorrow we'll have an answer to the great problems of the world, thought Charles Justiz.

John Glenn's flight did not provide the world with a cure to cancer, but it afforded Charles Justiz a glimpse of what the world could be. For his son Charlie, now nearly 10 years old, Mercury and the flight of *Friendship 7* revealed the true nature of the universe: so awesome, magnificent.

Charlie shook his head as the staccato images of Glenn, inside his capsule, flashed across the television screen. *What crummy video,* he thought. The onboard camera was recording images every five seconds, and that rendered the picture as jumpy as an old-time movie. Every five seconds Charlie saw Glenn's focused, intense face behind the helmet visor staring back at him. So cool! And, yet so far away!

During the mission his dad had an idea: "Let's see how high he really is going. Let's plot his flight."

Dad raced to the bookshelves and grabbed Volume E of the Encyclopedia Britannica. *Earth.*

"Grab a piece of paper," he told his son. Charlie did as asked. "Draw a large circle at the bottom of the page. We'll make this the Earth."

Charlie slowly sketched a huge sphere.

"Let's see, the Earth is 4,000 miles big [its equatorial diameter is actually 7,926 miles] in diameter, and he's going 200 miles [actually, an apogee of 162.4 miles]," Charles Justiz determined. "Therefore, let's draw one to scale."

Like the trajectory engineers at Mercury Control, father and son used dividers and a ruler to create a scale drawing of the Earth in relation to both the altitude of *Friendship 7* and the vastness of outer space.

Charlie studied the homemade drawing; against the huge panorama of space, Glenn's flight barely registered a bump on the graph.

Wow! I didn't think space was that close, Charlie thought. *It must be easy to get there.*

In time, Charlie Justiz learned that reaching outer space would be a lot easier if there weren't a nasty barrier called gravity. *If we didn't have to yank so hard on the rope to pull ourselves up.*

By the launch of *Friendship 7,* Charlie Justiz was entertaining ingenuous notions of following in his hero's footsteps. *To be able to climb that rope off the world . . .*

When Project Mercury concluded in 1963, he sensed NASA was burdened with its own internal "gravity." *They're not going to let a Cuban kid into space,* he feared.

It took him two years and a taste of flight to disabuse him of that self-limiting notion. In 1965, at age 13, Charlie Justiz joined the Civil Air Patrol and learned how to fly. How different life seemed from miles above the ground! So different that soon he adopted a new view on NASA: *They're going to have to tell me no.* By the time man walked on the Moon, Justiz was flying Cessna 150 single engine planes.

He still saw NASA as an organization with too few Hispanics, African Americans, or even women. Yet, he forged on. Recognizing that most of the astronauts had come from the military ranks, Justiz enrolled in the Air Force Academy in 1970, got his pilot training there, and graduated in 1974. In 1977, he attended test pilot school at Eglin Air Force Base in Fort Walton Beach, Florida. It was there that the Shuttle Approach and Landing Tests Program (ALT) was held.

Once Justiz saw that huge bird swooping down toward him, he was hooked. This Space Shuttle was a different vehicle. *It was a real spacecraft,* thought Justiz. The other programs consisted of *pods;* you popped them up and dropped them into the ocean, with limited control. The Shuttle had to be *flown,* brought home with the deft touch of a skilled surgeon performing delicate microsurgery. NASA was finally beginning to achieve its promise of easily accessing outer space.

☆ ☆ ☆

Two decades later, Charlie Justiz teaches the Shuttle astronauts how to land the swooping bird. A veteran of 17 years with NASA, Justiz has been a research pilot the last 15 years, instructing the astronauts on reentry and landing procedures.

It's a good job, he says, not bad as a second choice.

Yet, there is no doubt he'd prefer to be flying *Discovery* 150 miles above the Earth to a modified Gulfstream II a mere 14 miles above Houston.

"I would rather be driving the boat than watching it sail away," said Justiz.

He has interviewed for a spot in the astronaut corps four times. He has failed each time, most recently when he was medically disqualified because of a kidney stone. It's tough to take a B-57 high altitude reconnaissance airplane 70,000 feet and not yearn to kick it forward into the heavens. To see the black sky, the gentle gradation of the horizon. To realize how vulnerable our little spaceship is.

☆ ☆ ☆

To Dayna Steele, this was like Santa Claus canceling Christmas because of a snowstorm. Christmas Eve without luminarias? *Bah, Spacebug.*

It drove her crazy to think some of her neighbors knew nothing of the tradition, had no clue why each December 24, she and most of the residents of Seabrook, Texas, lined their yards and driveways with the festive lanterns. It shocked her to see young couples stare blankly as their neighbors purchased dozens of the holiday candles. *Each year, fewer and fewer people are taking part in the ceremony. I hate like hell to see it die,* she thought.

The lighting of the luminarias was a time-honored holiday tradition dating back to the Spanish conquistadors. The luminarias, or *farolitos,* as they were also known, were small votive candles burned in sand-filled paper bags. When lit, the candles glowed like cheerful, bright lanterns and turned the black night sky into a fountain of light. Historically, townspeople lit them to signify the coming of the Christ Child.

But on this Christmas Eve, Dayna Steele was not concerned about the religious implications of the ceremony. She was, as usual, on a mission to preserve and perpetuate the exploits of the astronauts, and luminarias played a role in her effort to do so.

Determined to have every family on her street participate in the ceremony, Steele purchased enough luminarias to decorate the entire block. Then, house by house, she approached her friends and neighbors.

"Will you buy the luminarias and light them?" she asked. "I'll even put them out myself and light them for you."

The people listened to her request; they perked an ear, raised an eyebrow when she accompanied her appeal with a story about a special Christmas Eve, 1968. The tale concerned three travelers who became purveyors of the Word. Their names were Borman, Lovell, and Anders, and from the heavens, they shared a message the world would never forget.

Steele completed the story, her audience rapt, silent.

"This is why we celebrate with the luminarias: to honor the accomplishments of Apollo 8, to remember the exploits of all astronauts from Shepard to Lovell."

That night, long after Steele had chronicled the mission of Apollo 8 for the final time, Seabrook, Texas was aglow in light cast by the lanterns. To Dayna Steele, the town had never looked so gorgeous.

☆ ☆ ☆

They soared above the most profound of altars, three messengers of God, drifting between the heavens and Earth.

Below them, God's landscape was so close, they could almost reach out and touch it; could almost feel the crater-filled, bumpy road of gray that unfurled like an interminable highway lined with countless rocks, all of different shades of gray. Except these stones, crags, and boulders were endemic to the Moon.

At 8:11 p.m., Houston time, on Christmas Eve, 1968, these three voyaging envoys prepared to impart the Word of God to an audience some 240,000 miles away.

Jim Lovell, Bill Anders, and Frank Borman had not traveled to a distant world simply to issue a long-distance Christmas blessing. They had flown faster, higher, and farther than any other human beings to help fulfill a promise made by a resolute president seven years earlier. The flight of Apollo 8 was the most ambitious of its kind: the crew ultimately would spend nearly two weeks in space on a journey that would bring the three men to within 100 miles of the lunar surface. The mission, which came less than two years after the tragic Apollo 1 fire, would be a giant step toward a lunar landing before the end of the 1960s.

Now, after circling the mysterious far side of the Moon, the astronauts gave a half billion people the television show of their lives: a tour of the Moon and its many landmarks, including the Sea of Tranquility, one day to be man's first intergalactic port of call.

Five hundred million wide-eyed Jay Egans and Scott Pelleys stared at their television screens as the astronauts transformed first into extraterrestrial tour guides and then into pseudo-clergymen.

One by one, each man reflected on his impressions of the Moon and the mission. Then, shortly before approaching a lunar sunrise, the astronauts offered a benediction from the Book of Genesis that would echo through the heavens for an eternity.

Anders began: *In the beginning, God created the Heaven and the Earth; and the Earth was without form and void, and darkness was upon the face of the deep; and the spirit of God moved upon the face of the waters.*

And God said, "Let there be light," and there was light.

And God saw the light, that it was good.

And God divided the light from the darkness.

Lovell followed: *And God called the light Day, and the darkness he called Night. And the evening and the morning were the first day.*

And God called the firmament Heaven. And the evening and the morning were the second day.

And God called the dry land Earth, and the gathering together of the waters He called seas. And God saw that was good.

☆ ☆ ☆

To a world suddenly so silent, so paradoxically distant, Borman concluded: "And from the crew of Apollo 8, we close with good night, good luck, a Merry Christmas, and God bless all of you, all of you on the good Earth."

As the crew bade their home planet farewell, Jim Lovell's neighbors in Timber Cove, Texas gazed up at the Moon as if they could somehow see the command module and its crew if they opened their eyes wide enough. Around them, the residential subdivision of some 100 homes had transformed into a fairy tale festival of lights.

Dr. Lawrence Chapman felt the goosebumps rise, a chill of camaraderie sweep over him, as he listened to Jim Lovell read from the Bible. This was something extraordinary, he conceded. It was more than just images of astronauts from a world away. This was Jim Lovell, his neighbor, his friend. Chapman had attended church with Lovell and Frank Borman. Borman was a man of devout faith, a lay reader during Mass. The kind of guy who, Chapman believed, thrived in a disciplined, structured environment.

Chapman had always found Jim and his wife Marilyn to be genial, warm people. He had felt that way about most of the second and third waves of astronauts, which included Lovell. Jeffrey, the Lovell's youngest child, had grown up with Nancy, the Chapman's second child. They had shared the same bassinet, they had car-pooled together during kindergarten and pre-K. Lovell, like Borman and many of the other astronauts, exuded a stoic calm; to Chapman, he seemed utterly unaffected by the trappings of fame. The same could not be said for all of Lovell's colleagues, Chapman decided.

There were two types of public figures, Chapman concluded: the White House political types who bent over backwards to befriend every voter in the nation; then there was the One-on-One Guy who craved *your* friendship, and was wise enough to recognize that no one person could be everybody's friend. Jim Lovell was that One-on-One Guy.

Except now, he was going one-on-one with half a billion viewers across the world. Chapman glanced around and saw a room full of partiers enthralled by the exploits of their neighbor. Outside, the streets of Timber Cove twinkled with the light cast by hundreds of holiday luminarias. Tonight, even the stars were jealous of the luminescence of these lanterns.

☆ ☆ ☆

Gladys Lancon and Edith Balfe had been in attendance the day the Timber Cove Ladies' Garden Club came up with the idea of lighting luminarias on Christmas Eve. Lancon, who had moved to Timber Cove with her husband nine years earlier, annually decorated her yard's lamplighters with patches of green and red cellophane. The subdivision had also done so, down on the boulevard. When the tall, gas-powered street lights buzzed on, the globes glowed in Christmas colors.

At the suggestion of another member, Lancon, Balfe, and the rest of the committee decided the luminarias would also enhance the festive appearance of the neighborhood. That this celebration would coincide with the flight of Apollo 8 was merely coincidental. Yet, this twist of irony would, for years to come, add fodder to the town's folklore.

For Edith Balfe, the flight of Apollo 8 was more than just another step forward in America's quest to beat the Russians to the Moon. Jim Lovell was her family's neighbor, their friend. He was not Jim Lovell, astronaut, but Jim, the regular guy who lugged his garbage out to the curb just like any other mere mortal. Jim, whose kids played with her kids. Jim, whose oldest daughter Susan was as much a part of her household as her own kids.

This was not a faceless pulp magazine space voyager; this was the devoted husband and father who remained unflappable through any crisis. He was not a hero, Balfe thought, but a decent man who remained cool and calm when his kids were misbehaving or when his wife, Marilyn, was up in arms about a matter.

Right now he was flying around in a vehicle not much larger than a refuse can, and that was thrilling.

With Jim Lovell racing around the heavens, and Timber Cove's first luminarias reflecting their light off the canal, Edith Balfe felt a kinship with her neighbors, with the world. She felt extraordinarily close to God.

Like her mom, 9-year-old Carol Balfe was friendly with the Lovells. She had practically grown up with them; in fact she was Susan Lovell's best friend. They were inseparable; they climbed trees together, dressed Barbie in a magnificent wedding gown; water-skied together. Carol often slept over Susan's house, and late at night, when the others were asleep, the girls would slip out the bedroom window onto the Lovells' roof to munch on the expensive canned almonds they had stolen from Marilyn Lovell's cache. Together, they peered down on the shimmering Timber Cove canal, then tossed nuts to a waiting squirrel.

Susan had often told Carol that she was embarrassed by the limelight, the publicity, the vulturish photographers and news-hounds who clamored about her home during her dad's missions. She just wanted to be a kid, wanted her father to be around.

Carol Balfe didn't know Jim Lovell very well, for he was often away, training. The two families had gone boating together, however, and, for many years, the Balfes attended the Lovells' Christmas Eve Party. What a treat: Marilyn Lovell's eggnog was superb!

Occasionally, during visits with Susan, Carol would get a glimpse of the man with the dry sense of humor. One Saturday, during the annual Army-Navy college football game, Lovell was bantering with a friend on the telephone. Each time Navy would score, Lovell, an old Navy man himself, would dial up his Army buddy and give him a good old Navy razzing. When Army scored, the Lovell's phone would similarly ring; this time, Lovell was on the receiving end of his buddy's good-natured harangue. This exchange went on for the duration of the game, and left Carol Balfe with an impression of Lovell she would never forget. In the end, he *was* just like everyone else. When he wasn't wearing a space suit.

So Christmas Eve, 1968, was a unique day for Carol Balfe. This was Santa's night, yet Carol recognized something important was happening. Her best friend's dad was in outer space; he had even said a prayer to the whole world. As Santa loaded up his sleigh, Carol Balfe gazed outside at the street and saw the pretty luminarias so bright and cheerful.

She thought, *This is so cool.*

☆ ☆ ☆

Like Dayna Steele, Carol Heidhausen, née Balfe, will always associate luminarias with the flight of Apollo 8. For many Timber Covers, it is easy to do so. In an area wherein about a third of the population comprises NASA employees and families, the sight of the candles glowing in the still night harkens images of space flight. Many of the residents who don't work for the space agency are employed in the oil industry. Still, a few others, such as Dr. Larry Chapman, relocated to Timber Cove for completely different reasons.

Chapman already was an established physician in the Galveston area when he, wife Marilyn, and son Chap moved to Timber Cove, in October, 1962. He had been drawn to the pastoral splendor of the area, the elms, the oaks, the pine trees, and to the soothing sight of the water, great for water skiing. It was the perfect place to raise a family. That NASA was building a new facility only expedited the development of this community and its environs. When Alan Shepard blasted the U.S. into the space race, Houston, Texas, was recognized primarily as a leader in the petrochemical industry. But that all changed when NASA announced in 1961 that it would build a new facility outside Clear Lake, some 20 miles southeast of downtown Houston. Located on a 1000-acre plot donated by Rice University, the Manned Spacecraft Center (later named the Lyndon B. Johnson Space Center) would be used to select and train astronauts, and would be involved in the development of spacecraft and associated systems for flight. It would be the successor to Virginia's Langley Research Center, whose facilities provided much of the Mercury astronauts' training.

Over the next decade, the pristine ranchlands would grow into quaint NASA communities such as Taylor Lake Village, El Lago, Clear Lake, and Seabrook.

The fluid expansion of the communities was a windfall to Chapman's practice, but it came with a cost to his free time. While NASA worked steadily to build the Manned Spacecraft Center,

Chapman toiled feverishly to stitch, splint, and salve the agency's many injured who flocked in and out of his Seabrook office like passengers at a bus depot. One day, he treated 28 MSC-related injuries.

And that was just a part of his daily routine. As Timber Cove's only residential physician (most of the astronauts and their families were treated by NASA doctors), and with the nearest hospital some 12 miles away, Dr. Lawrence Chapman had to don the hats of obstetrician, surgeon, and registered nurse. When his phone rang, Chapman grabbed his medical bag, hopped on his single-speed jalopy of a bicycle, and pedaled to the patient's house. On weekends he logged more miles on that bike than a pro cyclist in the Tour de France.

A life devoted to medicine left little time for a social life, and even less time to follow the development of the fledgling space program. That Gus Grissom and Wally Schirra lived practically across the street was of little significance to Chapman. Chapman saw the Original 7 space voyagers as distinctly aloof, private men, who stayed pretty much to themselves unless you were intimate friends of theirs. Ultimately, Chapman would have more intimate associations with the second and third waves of astronauts.

Chapman's lukewarm appraisal of Glenn and comrades did not come from a disdain for public figures nor from a disrespect of manned space flight. He was, in fact, a proponent of Project Mercury.

But Larry Chapman was, like the astronauts, an educated man, with a bachelor's degree in engineering from Rice University; he had also been an Air Force meteorologist. Essentially, when Dr. Lawrence Chapman waved hello to Gus Grissom or Wally Schirra, he felt he was acknowledging a peer, no better nor worse than himself.

Should any one of them request his medical assistance, however, Chapman would be on his bike and pedaling to their house quicker than you could get an astronaut to volunteer for a mission.

In a sense, Chapman liked the underdogs: the NASA men and women who worked behind the scenes to build the organization, to do what was necessary to send Grissom and colleagues into space. The technicians whose tired faces never graced the cover of *Life* magazine. These were the people of *character.* Most of them were Chapman's patients, as well.

One such patient was Jack Kinzler, one of the key technicians and engineers instrumental in the creation and development of the Mercury capsule. In 1958, he and a group of engineers collaborated to form the Space Task Group, to be headed by Bob Gilruth. Soon Kinzler went to work creating life-sized replica capsules, which were air dropped and tested for reliability and practicality. They were the forerunners of the production types manufactured by McDonnell Aircraft Corporation of St. Louis.

For a man who grew up dreaming about launching ballistics missiles, Kinzler was standing at the precipice of a new and exciting technology. He knew then that he was the luckiest man alive.

One day, NASA engineer Max Faget approached Kinzler regarding the design of the capsule's escape tower, which was equipped with a three-nozzle solid fuel rocket. Faget feared that if configured incorrectly, the nozzles would damage the capsule by shooting flames across its skin.

With drawing board and chalk in one hand, Kinzler went to work to solve the problem. Finally, he devised a configuration with a 20 degree calibration; this would provide as close to a perpendicular lift as possible, without compromising the integrity of the capsule.

For Kinzler, it was a way of vicariously living the life of an astronaut.

He got to know the Mercury 7 astronauts during their training at Langley. Kinzler's office was adjacent to theirs, and he quickly opined that the astronauts *were* heroes, men of unquestionable moral fiber, possessing the courage of pioneers to a new land.

On the day Alan Shepard helped America take its first small step from this safe world, Jack Kinzler stood near the blockhouse at the launch pad at Cape Canaveral. Now Technical Services Assistant to the director, Bob Gilruth, Kinzler was as thrilled as any person alive to be a part of the space program from its inception. He was, after all, a part of history in the making. But there were changes in the wind, positive changes. It was time for Jack Kinzler and NASA's advance team to head west, to Houston.

In February, 1962, the same month John Glenn orbited the Earth three times, Jack and Sylvia Kinzler spent a long weekend in Houston, selecting a plot of land upon which they would build their new home. Construction on the Manned Spacecraft Center was scheduled to begin in a couple of months, with the facility opening officially in another two years. In the interim, manufacturing and testing would be held at nearby Ellington Air Force Base.

Kinzler had recently visited the Houston environs and had found Timber Cove to his liking. Now, with his wife Sylvia at his side, there was no question that the clean, quiet, quaint development was perfect. The couple met with a contractor, settled on a lot, then returned to Virginia. Construction began in June, and in October, 1962, the Kinzlers moved into their single-story contemporary home.

It was a friendly neighborhood, and people were cordial, but not the kind who banged on your door and invited themselves in for coffee every afternoon at three. But they were there when you needed them. And how could anyone lament having Wally Schirra and Gus Grissom as neighbors? John Glenn and Scott Carpenter lived a few streets over, on a cul-de-sac at the end of a boat canal. To Sylvia Kinzler, these astronauts were every bit as polite and down to earth as the media had portrayed them to be.

Most Sundays Sylvia saw the Glenns at Webster Presbyterian Church, where the astronaut was an elder, and Annie, his wife, was the organist.

The Kinzlers led an idyllic yet often exciting life that paralleled a Mercury mission: there were periods of eager anticipation, breath holding, then the liftoff, then more breath holding, and finally a sigh of relief. And there was a sense of great accomplishment. For Jack Kinzler, every successful manned flight was a personal success. From 1958 to 1977, hardly a day passed when he wasn't contributing in some way to the development of a spacecraft. Each NASA program started with models that required wind-tunnel testing, and that involved Kinzler. His models shop grew from a one person fly-by-the-seat-of-your-pants affair into a full-blown team operation, 200 employees strong.

For 35 years, until his retirement in 1977, Jack Kinzler lived the storybook life of a space traveler.

The 1960s were a boon for Timber Cove and its neighboring communities. The second wave of astronauts brought Buzz Aldrin, Pete

Conrad, Jim Lovell, and others to Timber Cove. In 1969, Aldrin became the second man to set foot on another celestial body. During that trip, he carried with him communion elements—miniature vial and chalice—donated by Webster Presbyterian.

It seemed there was a welcome home party for every astronaut who returned safely from space. They were, after all, Pete, Jim, Buzz, John, Scott, Wally . . . All guys who took out their trash.

Before Project Mercury came to a conclusion, a handful of Timber Covers went to work ensuring the community would never forget the astronauts' exploits. With the help of the local civic group, a community pool was built, replicating the design of the Mercury capsule. It had the same funnel shape, with the slanted section of the upper body flaring straight out, like a real space capsule. Two steps were positioned at the rear of the "capsule," where swimmers could "ingress and egress." For many years after, Jack Kinzler would be a member of the pool maintenance team.

Today, more than a generation later, the Mercury pool preserves the image and accomplishments of the Original 7 astronauts. The Christmas Eve luminarias also keep the astronauts in the minds and hearts of the townspeople. What began as a way to spruce up the town has become so popular, the local telephone company notes the lighting time (usually 7 p.m.) in the front of its directory.

A week before Christmas, civic club members and other volunteers muster to coordinate the event. A truckload of sand is delivered to the local marina; it will be dispersed and allocated incrementally to volunteers who will package the luminaria kits. Group members then contact each resident to determine the quantity they will purchase. The Kinzlers usually light 60 lanterns each holiday. Volunteers set up shop in their homes, and residents typically pick up their allotment Christmas Eve day. With 100 percent town support, the process runs as smoothly as a typical Space Shuttle flight.

The Kinzler's corner lot provides an ideal setting to display the lanterns. Spaced some four to five feet apart, they transform a cozy home and the lush green yard into a giant Christmas tree.

Just before seven, the residents march out their front doors like synchronized soldiers on a mission. They wear wristwatches, hopefully synchronized with those of their neighbors. They bring lighters or matches that, hopefully, will work first try.

Finally, at 7 p.m., one by one, the delicate luminarias spark to life. A dancing tiny flame breaks the clear black sky, then another and another. Finally, all of Timber Cove is transformed into a magical fairy tale land of lights, flickering into the night. By this time, Sylvia Kinzler's arms and legs are rippled with goose bumps.

For most residents, Christmas Eve provides an opportunity to socialize with friends and family. For others, such as Jack Kinzler, it is a night to toast Jim Lovell, John Glenn, Gus Grissom, Bob Gilruth and the many NASA personnel who made space flight possible.

It is also a night to grow closer to God, as did so many people the night the Apollo 8 crew held the world spellbound with readings from the Scriptures.

Typically, the Kinzlers, and most Timber Covers, meander down to the small bridge by a boat canal that is just one giant leap from the Mercury 7 pool. They sing Christmas Carols, one and all, tenors, baritones, and frog-voiced prepubescents. A guitarist and other musicians provide music and harmonies. Then the residents gaze into the canal, bordered with luminarias, shimmering ghostlike in the night. Some close their eyes and recall the Balfe's 17-foot ski boat decorated with bright Christmas bulbs. If the weather is mild, as it usually is at Christmas time, most families stay out until 8:30 or nine. Then they take the short walk home, hypnotized by the luminarias' calming beauty.

Often, during this stroll, Jack Kinzler will gaze up at the Moon, the brightest luminaria of all, and mentally journey back to Langley, to MSC, to his model shops. And he'll feel like a kid once again.

☆ ☆ ☆

Hank Hartsfield was getting a haircut the day Alan Shepard changed the course of his life. To be more specific, he was listening attentively to flight updates on the radio as the barber's scissors clipped around his earlobes. Hartsfield became a space-a-holic that day. And his passion for space flight has not waned since.

Come Christmas Eve, Hartsfield has much for which he is thankful: Three Space Shuttle missions, a successful career with NASA. Everything he could have imagined that day, sitting in a cramped barber's chair that was not much smaller than the capsule in which Alan Shepard rode. Christmas Eve is always a special time around Hartsfield's home in El Lago: The kids come over, along with the grandkids. The luminarias make Christmas brighter. They help bring people together for a single cause, much the way a launch does.

"It has special meaning," said Hartsfield. "We do it every year because it's Christmas."

Mention to Hartsfield the relationship between the flight of Apollo 8 and the luminarias and Hartsfield raises an eyebrow: "That's interesting. I didn't know that."

☆ ☆ ☆

"October is not too soon to be thinking of luminarias," read the release in the October, 1997 edition of *La Ventana*, El Lago, Texas' community paper. "We will sell them again this year to support this publication and, of course, because they have such a lovely holiday tradition in El Lago . . ."

"They are sold for $9 in sets of 30 candles and bags. . . . The local tradition began in Timber Cove during the Apollo Program in 1968 when local astronauts were in space at Christmas, as a symbolic holiday greeting to them as they circled the Moon."

☆ ☆ ☆

The spirit of the message is appropriate; only the "fact" of the luminarias' origin in Timber Cove is slightly wrong. Today, most citizens of Timber Cove and El Lago associate the luminarias with spacebound neighbors, even if that slightly twists the truth.

Some folks read about the lore of Apollo and luminarias; some perpetuate the myth; others live it, particularly El Lago's historian, Glendora Hill.

Hill has participated in the Christmas Eve ritual for 15 years. Although the fund-raising event serves a practical purpose, Hill feels people should also remember and respect the accomplishments of the astronauts. It isn't always easy to do so, she concedes, when space flight has become almost routine.

That's not to say El Lago has completely turned a cold shoulder to the research done by Space Shuttle crews, or by space exploration in general. As president of the El Lago Astronaut Hall of Fame, Hill ensures that the images of Ed White and 45 other former and current El Lago-dwelling astronauts remain imprinted in the residents' minds. Forty-six portraits of 46 astronauts adorn the walls of City Hall. Forty-six unique stories that flash into the residents' minds whenever they walk through these halls.

Hill's job keeps her busy and in touch with the daily itinerary of NASA, for tomorrow's flight may take El Lago astronaut number 47 into low Earth orbit. That would necessitate yet another portrait for a wall already chock-full of history.

Like Sylvia Kinzler, Glendora Hill was one of the first to settle in a NASA community. She was living in Texas City, working in the petrochemical industry, the day of John Glenn's flight.

"When that happened, I called my mother right away, and said, 'Isn't it wonderful that we live in America?'" she recalled.

For many years, Hill lived five blocks up from Neil Armstrong, commander of Apollo XI, and the first man to walk on the Moon. A few years ago, she considered purchasing his home, but passed when she decided the floor plan didn't suit her needs.

Christmas Eve takes her back to the Glenn flight, back to Apollo XI, back to a time when it was easy to feel like a little girl on the night of Santa's arrival. Christmas Eve: a night when the Hall of Fame astronauts' portraits come to life, and the town buzzes with images of missions past, present, and future.

Most go this way: December 24, nightfall. *It's like Christmas in Maine, or a place where there's snow,* Glendora Hill thinks, wistfully.

Some 26 miles separate El Lago from Houston, but the worlds are as dissimilar as the North Pole and New York City. The constant, throbbing hubbub of Houston; the quieting beauty of El Lago. A lucky few know what it feels like to return home on Christmas Eve after shopping all day in Houston: the candles, glowing up and down the streets, serving as an illuminated runway into town. They say *welcome.* They warm you.

On the holiest of nights, Glendora Hill is taken back to her childhood, when sleigh bells really did ring, and the Christmas wreaths and trees reached out and grabbed you, hugged you. The night talked to you then.

Now it is so mighty, it feels so mighty.

Glendora Hill smiles to herself and reflects: *God is among us; yes, He is.*

The Forever Little Boy

Scott Pelley had not fulfilled his childhood dream of strapping on a space suit and voyaging to different worlds, but today came awfully close. It was September 29, 1988, and Pelley, a 31-year-old reporter with Dallas, Texas, ABC television affiliate WFAA, had joined dozens of fretful, shaky-kneed reporters to cover the launch of *Discovery,* STS-26. The launch was significant for two reasons: First, this was the first Space Shuttle mission since the *Challenger* tragedy two years earlier. Second, this marked the first time Pelley had covered a live launch as a reporter.

The space beat was not new to Pelley. He had covered previous flights, but never from the T-Minus stage of the mission. As he gazed across the lagoon at the immense machine that would vault the United States back into space, Pelley was swept up in a wave of sadness and sentiment. Born in 1957, Pelley had grown up with projects Mercury, Gemini, and Apollo; he had considered NASA and its astronauts indestructible machines, immortal spirits empowered to lift mankind to greater heights. Even during the near disaster of Apollo XIII, Pelley knew in his heart that somehow Flight Director Gene Kranz and his crew of technicians and engineers would get the crew home. Pelley felt that same vital force, that unshakable optimism whenever he watched a mission on television; as a reporter, his passion for space flight had not diminished.

But today, a seed of doubt lingered in his mind. He stared apprehensively at *Discovery* and wondered, *Will it blow up? Will it even work?*

Pelley was not being paranoid; he had witnessed the agony of the *Challenger* tragedy. His station had sent him to cover the story shortly after the explosion, and Pelley had found that the nation shared profound grief over the loss of the crew. For Pelley, it was like discovering your heroes were merely human, flawed, mortal.

"I guess we forgot why we launch over the ocean," Pelley told a colleague at the time.

As the countdown to the launch of *Discovery* ticked by interminably, Scott Pelley was flooded with memories of *Challenger*. He could still recall an interview with Senator John Glenn shortly after the 1986 tragedy, when Glenn said, "It was bound to happen . . . Given the energies and technologies and the complexities involved, this was going to happen, and the American public had become too complacent."

Maybe it was true, Pelley considered. Maybe America had taken space exploration for granted. He had not, certainly not as a child who had fashioned ersatz space capsules out of refrigerator boxes. Nor had Pelley, the reporter, been flippant about the space program, particularly when he had witnessed first-hand the many crucial components that had to work together just to make the huge machine lift three inches off the ground.

Scott Pelley was anything but cavalier right now. Despite his misgivings, he had not lost faith in the thousands of people who worked as one to make this day happen.

It has to happen, because these are our best and our brightest, he thought.

Discovery had to succeed, Pelley decided, because we had wrapped the space program in the nation's flag; moreover, we had already seen the stars and stripes ripped to shreds in one nightmarish split-second.

Amidst this torrent of images, thoughts, and feelings, Pelley turned his attention to Dick Covey and the crew of this mission. As a kid, Pelley had wondered how the astronauts had the guts to accept what many considered a suicide mission; now, as a reporter, he marveled at the mettle, the focus of today's team, which somehow had the courage to climb back into that cockpit after witnessing the shortcomings of technology—in the case of *Challenger,* the solid rocket boosters.

When *Discovery* finally rose tentatively to life, Pelley felt a profound connection with the hundreds of thousands of people who lined the nearby roads and marshes, watching as nervously as he. *It's their collective human will that is making this baby go, not the solid rocket boosters,* he thought.

Scott Pelley was a kid again, with his refrigerator box-rocket, thinking GO, GO, GO! *Everyone is wishing for the same thing, and it's a powerful wish.*

As the mighty rocket rose faster and faster and shrank smaller and smaller, the writer in Scott Pelley began to take over.

Our prayers rose with the Shuttle, Pelley thought. It would make for an effective lead in his story, he considered. A bit emotional, overwrought, perhaps, but right now his mind was being overpowered by his heart. People's prayers *were* rising with the spacecraft.

None more than his.

☆ ☆ ☆

The little boy spirit that had sustained Pelley through his coverage of the *Challenger* disaster was born about the time Gordon Cooper returned to Earth aboard *Faith 7.*

By the time Project Gemini was in full swing, Pelley was a full-fledged member of the Mercury Generation. *Nothing is going to stop me from being an astronaut,* he vowed.

Well, Pelley forgot about his mom, Wanda, who did not always appreciate her son's flights from reason, also known as Flights to Another World.

One day, seemingly like any other, she went to check on her eight-year-old son, who had been mysteriously quiet all morning.

"Scott, are you all right in there?" Wanda Pelley asked, hesitantly. She listened anxiously for her son to reply. A moment of silence passed, then another. "Scott, you have been in there all day; don't you have to go to the bathroom, or get a glass of water?"

Still, no response. She was beginning to worry now. *What possesses a boy to lock himself in his bedroom closet for an entire day? I hope he's not doing something he shouldn't be doing. He said something about a space mission . . . Mercury 13 or Gemini 2 or whatever . . .*

Wanda Pelley tapped on her son's closet door one more time.

"Scott, would you like some lunch?"

A few more moments of cold stillness elapsed. Wanda Pelley was about to give up, when finally the door opened slowly, just wide enough for Scott's head to emerge.

"Mom, I'm in isolation chamber training," Scott Pelley groaned, his face pinched with displeasure. "Your knocking on the door does not help me in this pursuit."

He shut the door and returned to his dark, claustrophobic world. *Mothers just don't understand what an astronaut has to go through,* he thought.

No, mothers were truly lost when it came to sensory deprivation, centrifuges, and a dizzying, death-defying machine named MASTIF. His mom just didn't get it when it came to standing still for 12 hours in a crushing room devoid of light and sound. John Glenn had done it, as had Wally Schirra and the other Original 7 astronauts.

Actually, they weren't in a closet, but in an enclosed chamber that afforded less comfort than Scott Pelley's ersatz isolation box. NASA's version was so dark you thought you saw phantoms; so quiet, you thought you heard the walls talk. NASA's scientists theorized that if an astronaut couldn't cut it in an isolation chamber, he couldn't bear the discomfort of a cramped Mercury or Gemini capsule during long-duration flights.

So that's why Scott was ticked off at his mom: She failed to respect his needs. After all, had John Glenn's mother offered her son breakfast during his second Earth orbit? Had Gordon Cooper's great aunt contacted him during his 15th orbit to ask if he had washed his hands after eating lunch?

No! It was so unfair that his mother didn't appreciate his efforts to prepare for his future. This was the 1960s, after all, and who knew how long it would be before he was training for a flight to the Moon? His mom should have understood what he was doing. *I'm not freaking out, after all,* he thought.

☆ ☆ ☆

Wanda Pelley should have been accustomed to her son's apparent flights from reason. If any kid in Lubbock, Texas seemed preordained to fly in space one day, it was her son Scott. From the first moment he watched a Mercury Redstone rocket blast off into space, he was hooked on space flight. By the advent of Project Gemini, Scott Pelley just *had* to be an astronaut.

His modest mechanical skills were augmented by an unflappable will to conquer space, even if only in his parents' front yard. Even if his homemade multistage rocket and capsule would reach an apogee of 3.2 inches. Scott Pelley followed the lead of his contemporaries of the Mercury Generation: he acquired a cardboard refrigerator box, then donned his engineer's cap. Inside the box, he used a marker to draw control panels, rocker and toggle switches, meters, knobs, and windows that would command him an outstanding view of the Moon. Then he went to bed, knowing he would never sleep, feeling as hyped-up as a rookie flyer the night before his maiden voyage. Adventure waited; there were all sorts of new worlds to conquer!

Long after most 8-year-old kids had fallen asleep on the latest issue of *Marvel Comics*, Scott Pelley tiptoed out of his room, then down the hall past his parents' bedroom. He collected his refrigerator-box-turned-replica-capsule and hauled it outside to Launch Pad 14, the front yard. *Darned hard work when there are no other technicians to help out!* He positioned the capsule on the lawn and aimed it toward the stars at just the proper vector. Then astronaut Scott Pelley strapped himself in and the countdown began.

Three, two, one . . . With an explosion, his rocket shook and rumbled with a thrust of some 50 pounds, about what he weighed dripping wet. *I'm off to discover other universes,* he thought.

Cape Pelley was the site of many impromptu liftoffs and splashdowns, and all were 100 percent successful.

Summers were the best seasons for blastoffs. Many nights, Pelley would climb onto the hood of his dad's Edsel to gaze at the stars, and ask himself: *What's out there? Why do the planets circle the stars? Will I ever get to one?*

Deep inside, Scott Pelley had a sense that he just wanted to *Go!*

☆ ☆ ☆

With an imagination as vivid as his, Scott Pelley couldn't possibly ignore the excitement, the drama of space exploration. Voyages to the stars were about all Scott Pelley thought of as a grammar school student: Manned flights, unmanned missions, astronauts, astronomy, the Moon, Mars. Science fiction books? He devoured them. Science fact? He memorized books and articles on anything pertaining to NASA and space flight.

While this time- and mind-consuming hobby kept Pelley from becoming a 9-year-old pirate, jewel smuggler, or transient book seller, Wanda and John Pelley were not overjoyed that their son was a space cadet. Wanda had only a passing interest in space flight; John Pelley was moderately interested in the accomplishments of the astronauts, but he was not a zealot of space exploration, as was his son. Still, he agreed to take Scott to Johnson Space Center the year man first walked on the Moon. During Apollo's heyday, he joined his son in a desperate search for a radio that was powerful enough to intercept the astronauts' communications with Mission Control. Dad and son never found such an instrument, but that failed to dampen the fun they shared.

Scott Pelley became a NASA junkie during the Project Gemini era. He marveled at every moment of every flight, hanging onto every description uttered by CBS newsman Walter Cronkite. Watching the missions, learning about the astronauts' lives, was, to Pelley, like being a New York Yankees' fan during their heyday in the late fifties, early sixties. It wasn't enough for Pelley just to watch his heroes on television; he wanted to be a part of the missions.

So Scott Pelley, utilizing a budding reporter's instincts and determination, wrote to NASA: "Could you send me photographs of the astronauts? Maybe even mission plans? Anything at all."

Someone, somewhere in the agency's public affairs office had a soft spot in their heart for Scott Pelley. Day after day, Pelley raced out to the mailbox. Much to his glee, he sometimes found a package boasting that cool NASA insignia. Inside, there were more delights than in Bozo the Clown's treasure chest. The neatest were shiny 8-by-10 color photos of Grissom and Schirra and the other guys. With a smile wrapping its way around his head, Pelley thumbtacked every picture, every mission plan, every snippet of NASA news and information to his bedroom walls. In time, his room would become a shrine to space exploration.

☆ ☆ ☆

As NASA graduated from one-man to two- and three-astronaut missions, Scott Pelley also advanced to a more technically comparable form of flight: model rocketry. This was the real thing: super cool rockets you prepared yourself, spiraling up into the sky, dropping softly back to Earth. During these model rocket missions, Scott Pelley became Chris Kraft, Mercury Flight Director. He had so many things to worry about in the days preceding a model rocket launch. Would the hardware work? The telemetry? Would he get a go from GUIDO, EECOM, CAPCOM?

Would it be Go, Go, Go from the "flight surgeon?" (God forbid he came down with a cold three days prior to a mission.)

Worst of all, would the weather hold up? *God, no rain, no rain!* That was the main thing; you just couldn't launch in inclement weather. Wind, too; you couldn't have too much wind, otherwise a gust might push your rocket into the trees, or out of sight, or, worse yet, onto the roof of your neighbor's house. Saturday was launch day; that meant Friday nights were pure hell, countless hours watching the weather reports or staring out the window into the sky. Rain meant sorrow, another long week of waiting, wondering and worrying like an expectant father. Sunny and mild was pure heaven. Cloudy with a chance of precipitation twisted Scott Pelley's stomach into knots: *Should I be excited?*

But the weather usually held out, and over the years Scott Pelley launched hundreds of model rockets.

He also watched countless hours of Cronkite, and watched Apollo XI in 1969. He was 12, and an avid photographer. That July night, Pelley, like countless others, took pictures of Neil Armstrong on the Moon, right off the television.

The impression of Neil Armstrong descending gingerly down the lunar module's ladder remained with Pelley throughout his youth and young adulthood. Furthermore, the magnificence of that mission encouraged him to cover the space beat when he became a reporter.

Now 41, Pelley has covered some 20 Space Shuttle missions, most of them as a chief correspondent with CBS News. Each and every one is as stirring as those fledgling Mercury and Gemini launches a generation before. Yet, Pelley has witnessed NASA and space flight evolve like a child into an adult. Orbiters capable of carrying a half-dozen astronauts into space have long since replaced the rudimentary, archaic one-man Mercury pods. Spacecraft no longer fall from the sky through a predetermined "window," to land in an ocean reasonably close to an aircraft carrier. Because of the Shuttle's broad cockpit windows, astronauts can truly see the "fireball" to which Glenn referred in 1962: the plasma gases shooting by, the immense wall of flame. Yesterday's splashdowns were heralded by the capsule's broad orange parachutes. Now, an Earth-shaking sonic boom announces the return of the giant V-shaped glider.

Today's technological advances sketch a mere outline of the story that is U.S. manned space flight. To find the heart, the theme of this tale, one must go back in time some 40 years. Then, Project Mercury opened the door of imagination through which walked Scott Pelley and many other members of the Mercury Generation.

"It gave us a fresh perspective on our world, knowing there is much more of life than what we see in front of us," said Pelley.

The Mercury guys are the pioneers to whom Scott Pelley refers when he begins wondering why the United States has a manned space program. Then he reminds himself, *We are now for the first time in our long history in a position to be able to go look in this neighborhood and step off the porch. We've always been drawn to the night sky; now's our opportunity to determine what is out there. We've stepped off the porch; there's a long way to go.*

As in 1965, Scott Pelley can look to the future of space exploration with a sense of optimism. His amazement with flight has been reborn in son Reese, five, and daughter Blair, two. Pelley considers himself fortunate, for Reese has a bent for space exploration and a keen interest in aviation and rockets. Thanks to his dad, he has a closet full of NASA toys, including two replica space suits, and a library of videos on space exploration.

Can he separate the science fiction of Star Trek *from the real life of the Shuttle?* Scott Pelley wonders. He giggles inside himself, recalling the time he came home with a videotape of the Apollo program, highlighting the lunar walks of Armstrong and Aldrin.

"Dad, did this really happen?" Reese Pelley asked his dad as they watched the retrospective together. Dad considered it incredible that in his lifetime he was able to separate science fact from fiction. "Yes, son, that really did happen," Scott Pelley answered.

☆ ☆ ☆

This is amazing, Shuttle *Discovery* astronaut Mark Lee thought during one of the few idle moments of his Extra Vehicular Activity. *How do I describe this view to the people back home?*

Back home was some 200 miles below him, a huge blue, tan, and white world as fragile as a newborn child, protected from the deadly vacuum of space by the thinnest film of atmosphere. *It's so round!* he thought.

And it was so quiet, so still; without wind as a source of friction, Mark Lee felt as if he and the spacecraft were standing still on a blanket of star-speckled black velvet. That was the illusion; both he and the Shuttle were speeding around the Earth at 17,500 miles per hour. With a slow blink of the eye, however, the coast of India hovered above him; another blink, and North America sprawled as far as Lee could see. This process occurred over and over again, every 90 minutes.

Discovery commanded an excellent view of the Earth, but to Lee it was like witnessing a sunny day through the window of an air conditioned office. Now, during this space walk, Lee enjoyed an unobstructed view, and he could spot the entire eastern and western coasts of Africa.

I'm almost on sensory overload. I just can't believe what I'm doing.

As the planet sped by in its paradoxical non-motion, it became an open book to its past, its present, its future: chapters filled with the calmest of the bluest oceans, the cheeriest puffy white clouds. Then there were the sobering chapters that spoke of the burning forests, the pollution, the oil-slicked bodies of water.

If only everyone was able to see the planet from this vantage point, maybe we'd have stricter rules for what we can do with our environment, he thought.

With a twinge of regret, Lee wondered: *Maybe we're just destroying this beautiful planet.*

☆ ☆ ☆

Three missions earlier in his career, Mark Lee might not have found the time to gaze wistfully at his distant home. The February, 1997 *Discovery* mission was the veteran's fourth, and he had changed since his 1989 rookie flight aboard *Atlantis.*

Talk about sensory overload!

None of his three subsequent missions matched the drama and significance of that first flight, when he was like an imaginative little boy trying to discern the mechanics of gravity. Thinking about a guy named Deke Slayton, who had grown up some 30 miles from Lee.

Mark Lee had always been captivated by the sight of a Shuttle liftoff. So much power and energy bowling you over. Then, as suddenly as the Shuttle sprang off the pad, it was a tiny speck in the sky. The crew, the machine, suddenly seemed so insignificant compared to the world around *you.* Then, a week later, crew and Shuttle were gliding back into *your* world, *your* life. *That they're instantly gone is what's so amazing.*

To Lee, watching a live launch was almost as thrilling as being a part of one.

Almost. Until he experienced space flight for the first time, in 1989 . . .

☆ ☆ ☆

Inside the orbiter's cabin, awaiting the launch of *Atlantis*, Mark Lee possessed that incredible energy he had felt rage from other shuttles during liftoff. Now, he was slightly apprehensive. *Will the flight go okay?* he wondered.

The waiting, the waiting, the waiting . . . strapped into his couch for hours . . . waiting. Wondering . . .

Finally, the waiting was over. Lee felt the spacecraft vibrate, then roar off the launch pad. Feeling the machine's harnessed power exploding all over made Mark Lee feel as if he could accomplish anything. Before he knew it, he was weightless. Before he knew it, he was watching yet another sunset, seeing a new cloud pattern roll toward him, seeing the stars develop into bright points of light so very quickly. So much quicker than on Earth.

Everything about Mark Lee's first flight in space seemed to flash by before his senses could assimilate the many profound sights with which they were bombarded. *Too fast* . . . When it was over, STS-30 was like a dream from which you awaken uncertain where you are, or if you were dreaming. The four-day mission flashed by so quickly, Lee hardly realized he had launched a satellite during the flight.

God, for John Glenn to orbit the Earth three times, it must have seemed like a half second to him, Lee thought.

In retrospect, space exploration had sent Mark Lee into sensory overload well before his first Space Shuttle mission. Lee's first taste of the anticipation and thrill of a manned space mission came 36 years before his most recent EVA.

Maybe Mrs. Parker feared the Soviet Union would drop a nuclear bomb on this one-room school house in Viroqua, Wisconsin. Maybe, Mark Lee considered, Mrs. Parker just thought these seven astronauts really were larger than life.

Whatever the reason, Mrs. Parker allowed Mark Lee and his classmates to watch the most exciting event of Lee's life: The blastoff of astronaut Alan Shepard. It would shape this eight-year-old's future.

If they let kids fly in space, he would have volunteered for the job. But the best that Mark and his buddies could do was ride with their imaginations to other galaxies with weird planets named Ziggleswoodleby or Star Base Yangle Dangledroid. They took these new worlds by storm, armed with the most powerful ray guns that could zap monsters with three heads, six eyes, and a peculiar ability to speak Peruvian French.

It was in that split-second, when Mark trained that electrified pistol on venomous demons, that he could see into the future . . . and visualize himself actually soaring out of his world. Maybe, one day, the aliens he dreamed of as a kid would really be out there.

Shepard's launch was not about chiller-killer laser beam guns that glowed red and fizzled and sizzled when you fired them. It was the real thing! That big, tall rocket with a real astronaut inside. Alan Shepard, the only person not relying on his dreams to fly him into space.

☆ ☆ ☆

Fifteen minutes were a lifetime for Alan Shepard and Mark Lee. By the time *Freedom 7* had splashed down, Shepard's world had changed. So had Lee's.

Shortly after the conclusion of the brief mission, Lee and a friend sat in the back of the classroom, still dreamy-eyed over Shepard's voyage.

"I have an idea," Lee said. "Let's get some clay and make clay rockets."

Good idea, agreed his pal. With Mrs. Parker out of the classroom, the two third grade would-be Mission Control engineers stole off with a huge ball of clay. Then, slowly, patiently, they molded the malleable potter's clay into crudely-formed Mercury capsules. To complete the process, the boys wet the "capsules," then prepared them for launch. *T-Minus seven . . .*

"Launch time," Lee cried, and the excited boys jumped to their feet.

SSSSSSHHHHBOOOM, they whistled through their front teeth, their chubby cheeks puffing and reddening like ripe apples. One by one the clay rockets leaped from the desks, soared into lower classroom orbit, then—*splat*—crashed into the ceiling, and hung precariously, as gravity worked to rip them back to Earth.

"Look at that!" Lee cackled, smiling at his accomplishment.

As they stared up at the ceiling, a shadow crept across the floor.

"Ahem," a voice said.

The boys' reverie was broken.

"Mrs. Parker . . . we . . . ah . . . we were . . ."

"Mark, why didn't you tell me what you were doing?" asked a put-out Mrs. Parker.

Lee pondered his dilemma; he was in deep trouble now.

"Well, we just thought they would stay up there a really long time," Lee responded, hoping his alibi would placate the teacher.

"Well, Mr. Lee, clearly you need a lesson in gravity. I want you to go to the library and take out a book on Sir Isaac Newton. Learn everything you can about the laws of gravity."

Mark Lee didn't just read the biography on Sir Isaac Newton, he devoured every page of it. A few days later, he stood before the class and, occasionally peering up to the spot where his clay models had once hung like mud icicles, explained how the laws of gravity worked.

While Alan Shepard's suborbital flight ushered the United States into the space race, it launched Mark Lee into a new and exciting world: the world of books.

After finishing Newton's life story, Lee returned to the library and took out another book. When he finished that one, he took out another, then another and another. Reading and learning became a compulsion for him. Between May and autumn, 1961, Mark Lee read nearly 200 books.

It would be one of the most significant events in his educational career.

☆ ☆ ☆

There was education derived through reading books; another type of schooling came while you were floating on your back in space, watching the sun rise on your home state. Or seeing the forests burn, the pollution rise like a giant nebula. Those moments floating silently in space were cathartic ones for Lee.

Everybody's got to do his part to take care of the environment, he thought.

Over the last eight years, Lee has done his part to protect the integrity of the planet by planting over 100,000 trees on his Wisconsin farm. The trees cycle carbon dioxide, thus making the air more breathable. In one respect, it's easier for Lee than for others to change the world; he has viewed the *whole* world, after all. In the blink of an eye.

☆ ☆ ☆

This was one of those days when Steve Smith was glad his parents were pack rats. The veteran of two Space Shuttle missions waited stoically for the 20 sets of eyes before him to blink. There was nothing unusual about this classroom of grade school students: Some wore quizzical looks, others stared absently into space. A few were entranced by the dog-eared drawings Smith held, at the head of the classroom.

"Did you really draw those when you were our age?" one student asked.

"I sure did, every one of them. I did them when I was in class," Smith added with a guilty smile.

The students stared silently. The many puzzled expressions told Smith the kids had not grasped the significance of the sketches, drawings of rockets, astronauts, capsules, and outer space, from a bygone era. Smith had drawn them in 1964, when he was in third grade, about the age of these kids. The illustrations had been preserved by his mother and father, who had refused to part with any of their son's creations.

"Some of those first astronauts were the ones who paved the way for my career as an astronaut," explained Smith, using his drawings as reference. "I never would have walked in space if it weren't for them."

"You walked in space?"

The inevitable avalanche of questions followed: "What's Mercury? Who's John Glenn? What's a Shuttle?"

"Let me tell you a story . . ."

☆ ☆ ☆

Steve Smith was born the year NASA was created, but he did not become interested in space exploration until Project Gemini was in full swing. Maybe his mom and dad didn't have a box at hand for him to convert into Gemini 8, but that didn't deter Smith and his buddies from playing Mission Control. Instead, when Mr. and Mrs. Smith weren't looking, Steve and his improvising fellow junior astronauts flipped the living room sofa upside down, converting it into a space capsule—dual hatches open, of course, for the impending space walks.

Then blastoff took place. Commander Smith ordered his crew to don the red Texaco fire helmets used exclusively during extended EVAs. One by one, they tethered outside their craft, creeping side by side past the end of the space ship that had the blunt heatshield. Only when the astronauts' life support systems began to fail—or Mrs. Smith summoned her son to the dinner table—did the missions conclude.

When not yawing and rolling in the living room, Smith adopted the role of flight director for his Mercury Astronaut GI Joe, and the extra special single-seat Mercury capsule. As he grew older, and Apollo succeeded Gemini, Smith turned his attention to a realistic three-foot model of the Saturn V. Anything that would connect him with space exploration.

Mom and dad played a role in Steve Smith's quest to become an astronaut. Bob Smith, an engineer at IBM, and Lilly Smith frequently took their son to San Jose International Airport where, for hours, they would sit in their car and watch planes take off and land. It was at the end of an airport runway that Steve Smith lost himself in the world of aviation.

He parlayed his interest in flight into an engineering degree from Stanford University. By his mid-twenties, Smith determined he had an aptitude for flying aircraft. He was selected by NASA as an astronaut candidate in 1992. Two years later, he served as mission specialist on *Endeavor* mission STS-68. And, some 30 years after wearing a Texaco fireman's helmet on his imaginary EVAs, Steve Smith performed three space walks as part of the STS-82 team that serviced the Hubbell Space Telescope.

Today, one of Smith's jobs is to help impressionable third graders appreciate the work done by astronauts. It isn't always easy.

Back when Grissom and John Young and Ed White rode atop Titan rockets, Steve Smith only had to pick up a crayon and sketch pad to become a part of their world. Television helped bring the missions back home; with just a few channels from which to choose, you watched the news, or a sitcom, or coverage of the current NASA mission. These days, however, kids could tune in to countless television stations and not hear a word about the latest mission. Computers and CD-ROMs virtually rendered sketch pads and markers archaic. *So many distractions, how do you get the kids interested in space exploration?* he wondered.

As a result, school visits are more than a brief show-and-tell; they provide Smith with a barometer of the new generation's interest in space flight. They would be incomplete without the typical Q-and-A period:

"What would you like to be when you grow up?" Smith asks the students.

He'll hear the same answers issued by his peers of a generation before: Soldier, doctor, football player . . .

Astronaut Smith will wait anxiously for the first boy or girl to shout: "test pilot" or "astronaut."

Instead, two little girls respond: "Veterinarian, marine biologist . . ."

Good, commendable professions. But . . .

Then a boy will join in: "I want to be Wayne Gretzky."

Doesn't anyone want to be the next John Glenn, Neil Armstrong . . . or Steve Smith?

Smith harbors no bitterness from the students' sometimes cool responses to his visits. After all, the world's political climate has changed; cooled, too. Back in 1963, it was America against Russia. Now, Smith watches kids discuss the recent 50-point effort by Michael Jordan, or the latest Nintendo game.

Thus, when lecturing, Smith has just a few minutes to become unforgettable, to make his mark, to say the right things that perhaps will lead a Johnny or Susie toward a career with NASA. He'll try to convince them that the Hubbell Telescope is a marvelous invention that can seize a person's imagination more than a three-point shot in the NBA finals. You can't see other worlds from a free throw line, he'll say. In space, there's no Comedy Network or Home Shopping Network.

Ultimately, however, Steve Smith's best sales tools are his space drawings, for they come from his heart and his imagination.

☆ ☆ ☆

Air Force Captain Rick Chiavetta stared at the future and was immediately transported to the past. Some 50 feet away, swathed in gold blankets and bearing a protective radiation shield, stood the spacecraft that would place Chiavetta's name in the history books: *Cassini*.

Tomorrow morning, October 15, 1997, this odd-looking cosmic explorer, with a body like that of a LEM and a top that resembled a giant satellite dish, would be hurled out into space to begin a six-year journey to Saturn and a subsequent four-year study of the distant planet.

As Chiavetta, the NASA Jet Propulsion Laboratory Launch Controller of this historic mission, peered through the protective glass partition in the spacecraft processing facility, he thought: *I'm so lucky to be a part of this. This is how the guys must have felt at the outset of Mercury and Gemini and Apollo.*

Chiavetta, 33, was born after the conclusion of Project Mercury and became seriously interested in space exploration during Skylab, the successor to Project Apollo. Although he came to know names such as Glenn, von Braun, and Redstone through history books, Chiavetta rued not being alive to watch Alan Shepard set the stage for man's greatest adventures.

Over the years, however, he had vowed to make space exploration a part of his children's lives. Remembering that promise, he looped an arm around the shoulder of Ali, his 4-year-old daughter.

"Honey, you're one of the last people ever to see this thing on the ground," he said, pointing to the enshrouded spacecraft.

Ali angled her head and scrunched up her face, as she tried to make sense of the weird machine with the pointed antenna on top.

"It's pretty," she responded simply.

"When you're a little older, this thing is going to be on Saturn, taking pictures and relaying information," her father boasted.

Chiavetta peered into his daughter's eyes. A small smile wrinkled her face. *It's neat,* her innocent eyes told him.

To help preserve this moment for the future, Chiavetta took pictures of his daughter and the Cassini spacecraft. He had already assembled a representative collection of mission paraphernalia; down the road, as the mission neared its completion, he would break out the memorabilia. Together, he and Ali would look back in time, to this night.

"Do you remember this?" he'll ask. "We were a part of that at the beginning."

At 4 a.m. the following day, Ali Chiavetta was joined by her 18-month-old brother Kyle, their dad, their mom Robin, and Dick and Mary Anne Chiavetta to watch the liftoff of *Cassini,* Rick's mission. Aside from the guys who created *Cassini,* Chiavetta was closer to this mission than anyone else. Although on launch day he used the call sign Launch Controller, Chiavetta's official position was Air Force Launch Crew Commander. His charter, over the past year, was to monitor the step-by-step assembly of the mighty Titan IV rocket that would launch the satellite on a path to the distant planet. This Titan was special: the first of its kind with a cryogenically fueled upper stage, versus the solid fuel upper stage of a typical Titan IV.

Like a mother-to-be monitoring the 40-week gestation of her child, Chiavetta monitored the arrival and subsequent testing of each rocket component. Slowly, piece by piece, these components were assembled into subsystems which, ultimately, became the mighty Titan.

Now, as the minutes counted down to seconds, Chiavetta felt the butterflies in his gut, the chills racing up and down his spine. The nights of waking up sweating over another problem that could jeopardize the mission were coming to an agonizing conclusion. With the countdown came new worries: *Is this thing going to make it off the pad? Will it veer off course and have to be blown up?*

Finally, the liftoff! *Go! Go! Go!* Rick Chiavetta shouted inside his head with such force it scattered the butterflies, warmed the chills, and silenced his internal Devil's Advocate. The Titan soared into the sky, taking with it the *child* in Chiavetta and a bit of the Launch Controller's soul.

Three-quarters of an hour after the launch, Rick Chiavetta was puffing on a celebratory cigar and toasting with a cup of beer, a home-made brew he had named "Cassini Launch Lager."

His job was over.

☆ ☆ ☆

In the 1960s, Allentown, Pennsylvania, was anything but a hotbed for space buffs. In fact, with Pocono Speedway a short drive away, kids like Rick Chiavetta grew up dreaming of racing cars and checkered flags. While members of the Mercury Generation fantasized about slipping into a Saturn V command module, Chiavetta longed to slip into the driver's seat of an Indy-type race car, like the one Al Unser drove: the *Johnny Lightning*. Dark blue, sleek, sporting a shocking lightning bolt down the side. So cool. So neat, in fact, that Chiavetta grew up dreaming not of being Neil Armstrong but of being Unser. The first Hot Wheels cars Chiavetta owned were replicas of Unser's car. When Chiavetta saw Unser endorsing Viceroy cigarettes in magazine advertisements, he announced to his father: "When I grow up, I'm gonna smoke Viceroys."

At age 10, Chiavetta was not about to take up cigarette smoking simply to emulate his hero, however. Neither could he drive a racing car at 150 miles per hour, although he fancied driving faster than the wind. But Rick Chiavetta could explore, in his own way, around the many trails near his home.

He and his buddies had a whole world before them to discover: a mountain to climb, one that ran from the end of their street to an old water-filled quarry they called The Mine Hole. It was no more than a mile in distance, but it snaked wildly, with more twists and turns than a roller coaster. There were dozens of offshoot trails from which they could choose; trails that might lead them to . . . who knows? Sometimes they ended up at Big Rock, a huge, house-sized pair of boulders on the mountainside. Some days, the gang took their dirt bikes. Other days, they brought lunches. Always, they were just looking to find something new, some trail, some canyon theretofore unexplored.

"Let's just go, go, go!"

No one great epiphany drew Rick Chiavetta toward space exploration. As he grew up and began to take note of the media's romance with space flight and astronauts, Chiavetta discussed the missions with his parents. Eventually, Dick and Mary Anne Chiavetta concluded their son had more than a passing interest in the exploits of NASA. One day, Dick Chiavetta had an idea: "Let's go to the hobby shop and see what you think."

And what Rick Chiavetta saw forever sold him on model rockets and space flight: Estes . . . The king of model rockets. Even the *Mosquito*, a basic rocket kit you could assemble in five minutes, was like a mighty Saturn V, soaring hundreds of miles into the sky. *How can such a little vehicle jump so high?* Chiavetta wondered. *What's it like to be inside there? So much vastness up there, so much sky . . . the sun, the Moon.*

☆ ☆ ☆

At age 12, Chiavetta was like the hooked NASA astronaut anxiously awaiting his next mission: "When are we going back to the model shop, dad?" he would ask.

Dick Chiavetta rarely announced the date of their next trip beforehand. But on those mornings when he surprised his son with a "Hey, son, let's go to the model shop," Rick Chiavetta could hardly contain himself.

"We going now, dad?"

"Not yet, son."

Five minutes later: "Now, dad?"

"Not yet, son," until finally, Dick Chiavetta relented.

Father and son made the short drive through Allentown, up and down a series of short side streets, until they arrived at the tiny, cramped model shop.

"Come on, dad," Rick Chiavetta shouted. Then he raced down the few steps that led into the shoe box-sized shop. Rick Chiavetta tore through the front door, his dad on his heels. He was oblivious to the models and the Cox airplanes on either wall. Straight ahead, on the pegboard at the end of the short hallway, hung clear plastic bags of Estes model rockets!

In leaping strides, the distance between him and that pegboard of rockets shrank and shrank like Apollo XI closing in on the Moon. *What models are in? Maybe* a *V-2,* or a *Thor Agena-B,* or *Astron Scout,* or *Little Joe II . . . Maybe the Saturn V? Oh, boy!*

Model rocketry became a family affair for Dick and Rick Chiavetta. They assembled the rockets in the family basement, then took them out to the fields beside St. John's Church.

On one of those special launch days, father and son were at their workstation, assembling another rocket. They were particularly meticulous with this one: Rick made sure the fins were exactly straight; dad spray-painted the rocket fire engine red.

"Let's take it out to the field," they shouted together.

Rick stationed the three-legged launcher at just the right spot, and aimed it slightly eastward to compensate for a moderate breeze. Then he positioned the missile on its launcher . . .

Three, two, one . . . swoosh . . .

A perfect launch! The red rocket soared skyward, followed by four wide eyes. At T-plus 33 seconds, the mission was optimum. But at the apex of the flight, something went wrong. Rick Chiavetta saw the *poof* of smoke that told him the parachute and wadding had been jettisoned from the engine and rocket body.

Something's falling, thought Chiavetta. *It must be the rocket body. The bungee cord that held the body must have severed or wasn't attached properly.*

Suddenly, Chiavetta's attention was diverted to a small device floating ever so slowly down and away from him and his father.

The nose cone and the parachute!

"Dad, run, it's floating away!"

Dick and Rick Chiavetta raced side by side through the fields, as if propelled by Saint John himself. But as fast as they were, the easterly wind continued pushing the nose cone and parachute farther and farther from their reach. It was heading down the block to their favorite pizzeria!

"Dad, it's heading toward Salvatore's!"

The tiny top of his new model rocket became smaller and smaller, until finally, it disappeared somewhere over the train yard.

"It's no use, dad, we'll never find it."

Rick Chiavetta now turned his attention to the missing rocket body. He had lost sight of it after the nose cone and parachute began their wayward flight. Trying to find this small conical device would be like trying to find water on the Moon.

So much for precision . . .

☆ ☆ ☆

It was the technology, the machinery, that attracted Rick Chiavetta to race cars, then model rockets, and eventually missiles, real rockets, and space flight. When not shooting a model rocket to Mars, Chiavetta picked up a three-eighths-inch wrench and helped out his dad with the family cars: the old blue wagon, the Falcon, the gray Plymouth, the green Pinto. Strictly maintenance stuff: oil changes, tire rotations, and tune-ups. But this hobby gave Chiavetta a taste of a possible vocation.

While he shoved a screwdriver into an engine block, Chiavetta also began to note the snappy, alluring ads run by the Air Force. After graduating from high school, Chiavetta enlisted, to pursue a career as a jet engine mechanic.

He spent four years at California's Vandenburg Air Force Base, working with Minuteman intercontinental missiles and Boeing engineers. Chiavetta's interest in mechanics blossomed significantly, thanks in part to his and his supervisor's mutual interest in machines and engines. The two would spend hours together, swapping engines or rebuilding the transmission of a car Chiavetta's boss had purchased. Finally, Chiavetta had his own project, a Dodge Demon, with 360 engine and 340 horsepower. Definitely a muscle car.

Although the engineering side of Chiavetta was beginning to grow, the explorer part of him still hungered to discover. While at Vandenburg, Chiavetta had two dirt bikes and a road bike. In a throwback to his childhood in Pennsylvania, Chiavetta and his Air Force colleagues often rode off on missions of exploration, through the canyons, the hills. *Just go, go, go . . .*

After Vandenburg, Chiavetta returned to Pennsylvania, convinced he could cut it as an aerospace engineer. He received his engineering degree from Penn State University - Harrisburg, then proceeded to Officer's Candidate School. Four months of undergraduate space training came next, then Chiavetta spent three-and-a-half years at Falcon Air Force Base before going to NASA.

☆ ☆ ☆

There are many perks to his job as Air Force Launch Crew Commander. The explorer in him enjoys a unique closeup view of liftoffs from the perimeter of the launch danger zone, some three miles from the rocket. As the countdown begins, he is alone with his thoughts.

I'm back at the beginning, one of those first lucky persons who stood here and watched Shepard and Glenn and the other Mercury astronauts leave our world. I feel good to have a connection with them. I guess we all came from the same roots.

As the powerful rocket leaps off the pad, Chiavetta is back in the fields with his father, watching the *Mosquito* leap over the trees, the slender rocket a mere speck in the sky after a few seconds. Then he is in the cockpit of the Shuttle, trying desperately to imagine how the G-forces feel, what the view is like. *I'd do this in a second if NASA asked me to.* His lungs cannot expand, will not allow air to flow in, until the Shuttle's solid rocket boosters detach and begin their descent.

Emotionally, the mission is only beginning. Now the worrying starts, the way it does for a mom whose child is heading off to her first day of kindergarten. *What will happen today? Will the flight be a success? Will the hardware work properly? Will the crew members be okay?*

If only he could be there, 116 miles into space.

Dealing with the seemingly endless journey of *Cassini* is even more daunting, for he knows he won't feel the completeness of the mission until 2004, when the spacecraft starts relaying data back from Saturn. *There's something out there . . . it's not nervousness, it's something out there until the mission is complete and a success. Cassini* is like waiting for someone to come back after taking a long vacation overseas. But eventually, this mission, like a Shuttle flight, will come to an end, just as a child's first nerve-wracking day of school draws to a happy conclusion. Chiavetta will be able to breathe again. *I don't have to worry anymore.*

He is, in part, engineer, pragmatist, mathematician. He is also that teenage explorer, spinning clouds of dusty dirt into the air as he spins out on a sheer trail. Rick Chiavetta's life is a balancing act, a constant left brain/right brain battle for control. To do his job successfully, he must remain centered. Ironically, to make it through some of the long days—some 14 and 16 hours long—it is the youthful *elan vital* that drives him.

There are times, however, when he permits the child in him free flight, particularly when he is with his children. Give Chiavetta a fleet of Matchbox cars or a train set, put him to work with his son Kyle, now 2½, and Rick Chiavetta is no longer 33 years old.

Somehow, Ali, now 5, seems predestined to a career in the aerospace business. Nine months after she stood hand in hand with her dad, trying to make sense out of that silly-looking *Cassini* satellite, Ali Chiavetta stood hand in hand with her father at Kennedy Space Center, where she and a dozen of her preschool friends were enjoying a tour of the facility's Saturn V Apollo Center.

"Ali, come here. Look at the Moon rock!" beamed her father.

Ali sidled over to the exhibit of the rock. She reached over and touched it, felt its bumpy surface. She smiled one of those ingenuous *I'm not quite sure what this means, but I know it's important* smiles.

Hours later, when they were back home, Chiavetta led his daughter outside and into the driveway. They gazed up at the stars, as they often did before bedtime.

"What do you see up in the sky tonight, Ali? There's the Moon, and you touched a piece of it today, didn't you?"

Ali looked up at the brightly shining Moon and grinned. This time, Chiavetta saw a modicum of understanding in her own shining eyes.

"Pretty cool, huh?" Chiavetta grinned.

"Yeeeeaaaahhhhh!"

☆ ☆ ☆

On the job, Chiavetta often returns to his youth. Sometimes this flight back in time occurs when he is giving a guided tour of NASA and the launch sites of the Mercury missions. "This is the spot where the first space flight in our country was," he explains. He peers into the faces of these visitors; he can see the awe in their eyes, he can feel their excitement. And through them, Rick Chiavetta is experiencing Alan Shepard's flight, for the first time . . . again.

☆ ☆ ☆

While NASA's primary goal in 1958 was to launch a man into space and return him safely to Earth, Vern Estes's chief mission that same year was to develop a self-contained motor that would allow rocketeers to launch a model rocket high into the sky and return it in one piece to the Earth, while allowing the rocketeer to retain10 fingers and two eyes.

Neither Estes nor the space agency had a simple task, for ballistic missiles such as the Redstone and Atlas, in their infancy, were as unpredictable as the miniature rockets launched by amateur rocketeers. The success of *Sputnik I* a year earlier had encouraged hobbyists to rush out and build amateur rockets with whatever components were available, often inappropriate or dangerous materials such as metal pipes, and hazardous chemical and explosive mixtures. The upshot of sloppy workmanship—and ignorance of the products involved—was often disastrous, and the industry buzzed with stories of individuals who lost eyes, extremities, even their lives, through amateur rocket explosions.

Clearly, a better method for building and launching commercial amateur rockets was needed. In 1958, Estes, then 28, operated a construction company and, like his parents, had a background in fireworks. His entry into model rocketry came at the request of G.

Harry Stine, rocket engineer and president of Model Missiles, Inc., of Denver, Colorado. A year earlier, Stine and Orville H. Carlisle had teamed with financial backers to form the company whose charter was to market a toy rocket system invented by Carlisle. In 1954, Carlisle's Rock-A-Chute motor and toy rocket gave model rocketry its modest beginning. The Rock-A-Chute was innovative because, for the first time, users could purchase a hand-loaded motor, thus virtually eliminating the dangers posed by amateur rockets. Stine and Carlisle turned to Lawrence Brown, president of Brown Manufacturing Company and an experienced entrepreneur in the fireworks market, to build the motors.

But by May, 1958, inconsistencies in the Brown motor forced Stine to seek out another manufacturer of his rocket motors. His search ultimately led him to Denver's Mile High Fireworks, owned and operated by Mr. and Mrs. Earl Estes, parents of Vern Estes. While they had no interest in entering the rocket motor industry, they suggested that Stine contact Vern, who had experience in manufacturing pyrotechnic fountains and other types of fireworks.

"We need a dependable motor, and we need someone to set up facilities to manufacture these small rocket motors," Stine explained to Estes.

Vern Estes considered Stine's offer and, a few weeks later, accepted the job. Estes stipulated that he needed time to design and build a motor manufacturing machine. Working in an old construction shed with an equally old lathe, Estes built the machine designed to produce a model rocket motor every 5½ seconds.

Theoretically, it was a simple procedure: load the component materials, turn the machine on, and—BANG, BANG—one after another the motors would pop out. But reality and theory did not always merge, and because of all the safety stops built into it, the machine would one moment run smoothly, then suddenly shut off. The fickleness of the automatic motor-producing device prompted an associate to nickname it Mable, after an equally fickle woman he knew.

On January, 15, 1959, the first production model rocket engine from Model Missiles Inc. was launched. Ironically, as inconsistent as Mable was at the outset, it soon began producing motors quicker than Model Missiles Inc. could sell them, for Stine had sold his product exclusively to hobby shops. Estes, whose contract permitted him to market the motors by mail order, began advertising his motors in mid-1960. While maintaining the construction firm, he sold the rocket motors in his spare time. But as the space race developed, and consumers became more aware of safer model rockets, Estes was deluged with orders. In August, 1961, Estes sold the construction company and moved Estes Industries to Penrose, Colorado.

Those early days were for Vern Estes as they were for Bob Gilruth, Chris Kraft, and NASA's other early space technicians: filled with excitement, promise, wonder, worry, and dreams for the future. Estes's manufacturing concept for rocket kits was simple: build a lightweight kit made of balsa wood or plastic and lightweight materials; and provide an electric igniter for launching, and, of course, pre-loaded (black powder-filled cardboard) motors, some $\frac{5}{8}$ inch in diameter and three inches long. His production methods would make rocketeering easy and safe for kids ages 10 to 100.

As the space race heated up, so did model rocketry. In 1961, Estes Inc. manufactured its first kit, the *Scout*, which employed a tumble method of recovery. That same year, Estes published his company's first product catalog and an industry newsletter called *Model Rocket News,* which informed Estes's customers of the advances in model rocketry. The publication also provided accounts of hobbyists' experiences with model rockets. Some anecdotes were so fantastic, they seemed borrowed from the pages of *Amazing Tales.* These articles reflected the bizarre measures some rocketeers would undertake to duplicate the feats of NASA. Many of the intriguing accounts concerned Estes' *Bug-A-Bye* payload rocket.

The *Bug-A-Bye* was designed specifically to carry a payload into inner space, anything from a small transmitter and camera to insects. During a launch with a series two engine, the payload would be subjected to a whopping 100 Gs. Under the same force, a man would weigh 17,000 pounds.

The results of such "manned model rocket missions" were often amusing. Consider the following article in the second edition of *Model Rocket News:*

Cricketnaut Launched

A young scientist, 14-year-old Bill Waldron (NAR #1014) of 2343 Raleigh St., Denver, Colorado, will go down in history as the pioneer of studying the effect of acceleration and rocket flights on the cricketnaut.

Recently, Bill, a member of the Mile High Section, brought two crickets out to the Hog Back Rocket Range, one named Leika and the other called No-Leika. These two untamed wild black beasts had been carefully prepared and trained for their place in history by being removed from their natural habitat and carefully concealed in a ventilated coffee can for four hours.

Upon arriving at the range, Bill and his medical and launching crew methodically went to work in a manner which would make the boys down at the Cape look like amateurs.

Their specially constructed rocket was brought out and carefully positioned on the launching pad. The countdown was started at X minus 15 minutes. At approximately X minus 12 minutes, final connections to the rocket firing panel had been completed and it was announced that the space vehicle was ready for launching. All that remained was the final check-out of the two cricketnauts.

The medical crew went to work. While no heart beat or respiration was detected, the crickets were assumed to be alive and healthy because they were moving about and chirping. The final decision of the medical men was that the two crickets were in perfect health and ready to enter the capsule in the nose of the rocket.

At approximately X minus 3 minutes, the crickets were carefully brought out to the launching pad and shoved into their scientifically designed compartments.

No-Leika was positioned in the upper compartment, which was not padded. Leika was placed in a more comfortable padded compartment with windows for his viewing of the Earth below.

Complete details of the rocket ship used in this experiment, its weight, size, payload compartment, etc., are still classified. However, we were able to learn that a #16 engine was being used. Considering the probable weight of the rocket, this would subject the two cricketnauts to around 50 to 100 Gs acceleration. Would they survive the tremendous force?

Time is running, X minus 60 seconds came over the communications network.

"Tracking East is ready."

"Tracking West is ready."

Final safety clearance is given for the launching. Time is X minus 5 seconds,

4—3—2—1—zero.

A loud cheer is heard from the members of the crew and spectators as the rocket zooms skyward under the tremendous acceleration. As the rocket reached the apex of its flight, the parachute was automatically deployed. (Due in part to lack of training and in part to doubtful aptitude and ability of the cricketnauts, it was decided to leave no important phase of the operation to their manual control.) Then began the long and tedious job of tracking the rocket on its descent to Earth. (Approximately 30 seconds.) Did the crickets survive? Would the padding help Leika? Would No-Leika survive without padding? It had been agreed in advance that only the medical team would be permitted to examine and interview the space heroes.

The capsule was down and the medical team rushed to retrieve it. No-Leika was unconscious. Oxygen was needed. Everything had been considered before launching. No detail had been overlooked. Oxygen was supplied in no time by fanning the cricket with a piece of cardboard. But still No-Leika lay on his (maybe her) back unconscious. Maybe he was dead. As before the launch, no heartbeat or respiration could be detected by the medical crew. The quick-thinking medical men called for ice water to revive No-Leika. Having no water readily available, a quick decision was reached to douse No-Leika with Pepsi Cola. It worked!! A cheer arose from the spectators as No-Leika sprang to life. What could have been a complete failure was now proclaimed a success.

Leika was found alive, chirping, and in the best of health. Apparently, she (or he) had suffered no bad effects of any sort. Perhaps it was because of the padding. Maybe the windows helped. Maybe Leika was simply a better cricketnaut.

Final Summation

In summarizing the results of the above test, we believe it would be safe to say that No-Leika no liked her rocket

> trip and Leika liked her rocket trip. But if Leika had been No-Leika and if No-Leika had been Leika, then Leika would have No Liked and No-Leika would have . . . Oh well, . . . Forget it.
>
> The above experiment is true. Only the names of the cricketnauts have been changed to protect the innocent. Also, a few details may have been altered or invented to suit the fancy of the author.

Clearly, a NASA centrifuge may have helped No-Leika better withstand the enormous forces she/he experienced during this "long-duration" flight. But even in this daunting sub-sub-orbital model rocket mission, cricketnauts fared better with Mission Control specialist Bill Waldron at the helm than did their namesake during its sole flight in space.

A month after shocking the world with the successful launch of *Sputnik I*, the Soviet Union was back at it. On November 3, 1957, a Russian rocket carried a payload into outer space: a dog named Laika. The pooch, a spirited little fellow, lounged in a specially-designed dog capsule. Apparently, Laika, like his future distant relative No-Leika, no-liked the mission. After a week of surviving without his dog bones, Laika died of asphyxiation.

If Project Mercury was a test of individual and national will and engineering savvy and perseverance, Project Model Rocket was a challenge of imagination. While rocketeers could safely launch their rockets without fear of hurling a thumb to Mars, they were faced with a whole new set of challenges.

First there was the hardware, namely the model rocket itself. One enterprising group of rocketeers built a wind tunnel to test the stability of the rocket in flight. They fashioned the device from a large squirrel cage reinforced with wood. The hand-powered wind tunnel, when cranked by a brawny person, could produce a wind velocity of 15 feet per second, enough to determine the stability of the rocket. These junior von Brauns now had a means of eliminating the potential for dangerous flights. No 8-year-old wished to see his rocket U-turn into the ground at 2,000 feet like an erratic Atlas.

Far more difficult than proving the model rocket flightworthy was choosing a suitable cargo for the rocket's payload. NASA had considered everyone from circus performers to race car drivers before finally choosing military pilots to man the first Mercury flights. Unfortunately, Project Model Rocket was without precedent; rocket engineers would be forced to rely on a trial and error approach to determine which insects, mice, or instruments were most suitable for flight.

If *Sigma 7* was cramped, then a two-inch model rocket payload section precluded the use of most guinea pigs, moles, bats, and chickadees. Leika and No-Leika were the ideal size, but was it prudent for Bill Waldron to launch a Gemini-type two-cricket mission prior to an experimental single-bug flight? Such were the many questions facing the Waldrons of the world.

Regarding the training of the astronauts, NASA had it comparatively easy. You could take Gus Grissom, spin him dizzily in a centrifuge, and determine by his reaction how he fared during the training. It wasn't quite as easy to train a cricket for 100 times the force of gravity. Who had a centrifuge that small? Moreover, for an 11-year-old kid with a 33-family paper route, the cost of purchasing some modest facsimile of a mini-centrifuge was exorbitant. And no one was going to ask his mom for an advance on his allowance just to buy an ersatz isolation chamber.

Project Model Rocket physicians had their work cut out for them. It was extremely difficult to determine the flightworthiness of a mouse if he would not turn his head and cough. You couldn't check for an irregular heartbeat, for no one had an EKG machine that small or sensitive. Ultimately, if a locust walked up and volunteered for flight, all a Bill Waldron could do was ask the willing pilot to sign some sort of release, in the likely event of its death.

Fortunately for model rocketeers, landings were completed over land—like the Soviet Union had done—and not on water. Land returns eliminated the need for aquatic rescue teams, such as a Labrador retriever. Also, since no one had developed a motion sickness bag for a centipede, cleanup would have been messy if rocketeers used the neighbor's pool as their personal Atlantic Ocean.

Despite these minor inconveniences and barriers, payload flights generally fared well. Occasionally, however, a tragic story would reach the copy writers at *Model Rocket News*. Consider the flights of two 'Mousetronauts' in 1962:

> *Mousetronauts perish in separate flights*
>
> Two recent attempts to send Mousetronauts, sealed within special space capsules, into inner space resulted in death for two space pilots when parachutes failed to return them safely to Earth. The two Mousetronauts were killed in separate accidents at Lititz, Pennsylvania and Puyallup, Washington when attempting to test new recovery methods.
>
> Officials in Washington, D.C., have not made comment about the possibility of sabotage, but the similarity of the time and causes of the two malfunctions should undoubtedly create official investigations.
>
> LITITZ, PENN. (MRN)—Young scientists, in charge of the recent space flight, reported that "Aries," the Mousetronaut who had previously made a very successful flight into inner space, was killed instantly when his space capsule failed to detach from the rocket during a routine flight earlier this year. Spectators reported that the space capsule was put into flight by a secret experimental two-engine booster. The main rocket was powered by a single engine. The scientists, Richard Gundrum and Carl Shenk, Jr., did not comment on the cause or causes of the ill-fated flight. The passing of "Aries" was a hard blow to the people of Lititz, as "Aries," a life-long resident of Lititz, had been the first Mousetronaut to make a successful flight over the state of Pennsylvania.
>
> PUYALLUP, WASHINGTON (MRN)— Tight security placed upon this incident has seriously restricted news coverage, but our correspondent, Steven Boone, reports that on January 13th, Mousetronaut "X," a resident of a small field near Puyallup, was killed in an unsuccessful effort to return from space when his parachute recovery system did not function properly. All that is now known is that "X" died when his rocket, with capsule still attached, came to an earth-shaking crash that shook the community of Puyallup.
>
> (Officials at the Estes Industries Plant at Penrose, Colorado, manufacturers of the rocket engines that supplied the power for the two separate flights, when questioned about the cause of the accidents, would only say, "The nose cones must have been too tight.")

There was both an ingenuous, light-hearted side to model rocketry and a serious and educational side. Estes rockets were used by science clubs, students at science fairs, and by Boy Scouts. Estes Industries helped promote the hobby by selecting a rocketeer of the month and by frequently promoting competitions. Also, The National Association of Rocketry annual meets gave serious rocketeers the opportunity to compete against their colleagues.

While NASA had the publicity machine to create interest in what it was doing, model rocketry relied primarily on word of mouth to reach prospective hobbyists. Not surprisingly, between 1961 and 1963, sales soared with the flights of Project Mercury.

Early on, Vern Estes recognized that the best part of model rocketry was the flight itself. To build the kit yourself, prep it, then launch it successfully, made little boys and girls dance around rolling fields exclaiming, "Look, Daddy, I did it."

In exchange for the thrill of actually completing a launch, those gleeful kids (and kids at heart) wrote to Estes: "Model rocketry has given me a direction, a career to aim for," some letters read. Others: "It has given me a purpose in life, made me interested in math and engineering. Maybe someday I will design a rocket." And, of course, the inevitable: "Thanks to model rockets, I want to be an astronaut when I grow up."

Over the years, Estes made it a rule to read every letter possible. Every comment, suggestion, and criticism was taken to heart. The company followed the trends of rocketry, and especially NASA, by launching a Mercury capsule in 1963, Gemini-Titan in 1966, and the classics, Mars-Lander, Orbital Transport, Mercury Redstone, and Saturn V, in 1969.

1969 was a significant year for Vern Estes. Late that year, he sold the company to the Damon Corp., a conglomerate that purchased a number of small hobby companies, including Centuri Engineering, another manufacturer of model rockets. In 1990, Estes Inc. was sold to Peter W. Smith and Company, and Hobby Products. Since then, Estes Inc. has acquired other companies, and today is the leading producer of flying model rockets, engines, and accessories, in the world, as well as a leading manufacturer of model airplanes, kits, flying helicopters and small engines for model airplanes. You can't walk through a hobby shop without seeing the Estes name.

Forty years after building Mable, Vern Estes still gets goosebumps whenever he sees a model rocket soar through a bright blue sky, just as he does when he sees a Space Shuttle blast off. He has been active in both worlds: the literal world of NASA, the creative, imaginative world of model rocketry. At age 68, he still yearns for both, and wonders why he ever sold the company that afforded him the opportunity to give small and big kids so much pleasure.

"I don't know if it's us keeping a little bit of boy in them forever, or whether it's providing a place for that little boy to express itself," said Estes. "It might have been there already; we all have a little bit of child in all of us."

☆ ☆ ☆

I*'m 12 years old, and what do I want for Christmas?* 40-year-old Matt Steele wondered. He probed the open faces of the 12 Cub Scouts before him. *What would I want for Christmas if I were them?* he asked himself. He knew how he would answer the question, were he one of these 10- or 11- or 12-year-old boys: *I want a 1969 Saturn V Estes Model Rocket.*

To truly understand what toys, gadgets, and electronic games interested today's youth, one had to think like a kid, *become a kid.* That was a simple task for Steele; he was not only the leader of this Cub Scout troop, he was a Marketing Coordinator with Estes Industries and Manager of the North Coast Rocketry brand of Estes. Ultimately, Matt Steele made a living playing with toys.

"OK, guys, Christmas is just around the corner. Who wants what?"

Steele expected to hear *Nintendo* or *Sega.*

"I want Sega!" squealed one toothy kid.

"I want a Nintendo!" chimed in another boy.

Nintendo, Sega, Nintendo, Star Wars, Nintendo . . .

No surprises here.

Finally . . . "I want a model rocket!"

Now you're talking, thought Steele. *What kind?*

A moment later: "I want a model rocket, too."

When the votes were all in, Steele counted three tallies for model rockets, nine for Sega and Nintendo. Pretty much what he expected: a 25 percent hit rate for the model rockets. Back in 1969 and 1970, when man walked on the Moon, an Estes Model rocket was, for a 12-year-old, a walk on the Moon itself. Word of mouth spread quickly, and when the company released models such as the Saturn V, you could bet most kids wanted that for Christmas.

While Estes Industries paid Steele to fiddle around with its rockets, his primary responsibility was to determine just what models the company should manufacture. That's why many of these weekly Cub Scout meetings opened with a brief question and answer session on model rockets. What better way to determine what kinds of model rockets kids were interested in than to ask the consumers themselves?

"So most of you guys want Sega or Nintendo. Anyone interested in Star Wars stuff or GI Joes or Power Rangers?"

"Oh, sure, but Nintendo's the best."

"What about models or video games?"

"They're OK."

Steele heard lots of "OKs", but few "Oh, yeahs!" . . . except when it came to Sega and Nintendo. But that was to be expected; these giants of the toy world had bottomless advertising budgets which allowed them to penetrate every conceivable demographic. For 40 years, Estes Industries had relied primarily on word of mouth, some magazine advertising, and the hobby newsletters to promote their products.

But working for a smaller company had its benefits. It allowed Matt Steele to stay close to the pulse of the industry; and every time he introduced this hobby to a nine-year-old girl or her 59-year-old grandpa, Matt Steele journeyed back to age 7, when he had bought his first model rocket.

The best way for Steele to interact with hobbyists and would-be hobbyists was through product testing. Before a new Estes rocket went into production, it was first field tested, usually by Steele and one other associate. Steele would visit kids at a school, club, or Boy Scout meeting. He would watch them assemble the kits, study the instructions, scrutinize the components, and, ultimately, fly the rockets. A child's smile was a staunch endorsement of a new design.

Steele was paid by the smiles he could elicit, but within those smiles, he claimed his ultimate reward.

One such smile remains with him each time he goes into the field for product testing.

☆ ☆ ☆

They're not into this, Steele concluded. He stared ruefully at a few of the boys who had splintered from their classmates. While the others eagerly awaited their chance to try their hands at assembling their model rockets, these wayward kids ran around aimlessly, horsing around, jumping on each other, and essentially giving Steele the cold shoulder. As they drew closer, Steele saw in their faces utter disinterest; their eyes were dull, as if they had not beheld magic in a lifetime. One of the boys, about age 10, was particularly aloof. *Maybe he's just one of those kids who is better with his hands than with a textbook,* Steele thought. *I gotta get into this kid's head.*

On this day, Steele was testing a simple model called *The Generic*, a rocket with an appropriate moniker. Plain and white, it came to life through a rocketeer's imagination. By applying markers and decals to their rockets, hobbyists could make their versions of *The Generic* unique. With an eye trained on the wandering boys, Steele began unpacking the rockets.

BOOM!

It was as if a Martian had suddenly appeared at second base of Yankee Stadium.

Suddenly those once-unconcerned, dull-eyed boys were flocking around Steele, pawing at the rockets. And that one apathetic youngster dug right in. He meticulously applied the decals and markers to his model, turning and twisting it—"Am I doing this right?"—and scrutinizing every inch of its body—"Is this right?"—as if it were a Saturn V hurling three men into the vacuum of space.

Steele watched with interest as the boy grabbed his rocket and raced over to the launcher.

'TEN, NINE, EIGHT . . ." he shouted in the loudest of voices. "SEVEN, SIX, FIVE, FOUR, THREE, TWO, ONE . . . BLASTOFF!!!"

The rocket leaped off the launcher and soared straight as an arrow into the sky. *So perfect a launch*, thought Steele.

"Give me a high-five!" the boy yelled to his buddy. High-fives flew all over the field, just as the rocket began its descent.

Then the boy turned to Steele and flashed him the biggest smile Steele could imagine. Steele looked into those once lifeless eyes. Now they were brighter than a thousand Shuttle launches: *Look what I've done!* they said.

Steele understood the nature of that smile, the power it beheld. He'd grinned like that a thousand times since age 7, grinned like that with a thousand liftoffs. Steele didn't know if this one simple launch—a great feat for this youngster—would solve every math or science riddle the boy would encounter during the balance of his schooling. But Steele recognized that a plain little model rocket had given this kid a sense of hope. *He's finally experiencing that feeling of accomplishment that no one can ever take away from him.*

☆ ☆ ☆

"Sit down," Grandma Emma ordered 5-year-old Matt Steele. The boy was wise enough to obey his grandma when she spoke; she was the visionary of the family, after all. The only relative he knew who had ever flown in an airplane, she had a penchant for distinguishing the important from the trivial. She'd witnessed the evolution of aviation; now she was a spectator to the dawning of a new era in technology—manned space flight.

"Today, Matt, something very important is going to happen," she told him. "John Glenn is going to fly in outer space. This is going to be a big thing, a part of history. You're going to want to remember this later."

Matt Steele sat quietly and watched the TV, saw the big rocket sitting there. What did it all mean? Suddenly, someone, something—a *spaceman?*—emerged from a big van. *This is incredible!* The spaceman who, as the TV man said, was named John Glenn, began to walk away. Then Matt Steele could see just that big silver and white rocket sitting there quietly.

Grandma Emma didn't bother to explain the risks facing America's newest hero. Five-year-olds knew nothing about orbits and weightlessness and rockets blowing up. Nor did they care. But when she looked into her grandson's eyes, she could tell something was going through his mind.

He's going on an adventure, was what Matt Steele thought. He sat erect through the countdown, as the smoke began spewing out from the rocket. Suddenly, that shiny rocket was jumping up into the sky, getting smaller and smaller, until Matt Steele saw nothing but gray sky on his grandma's black and white television.

☆ ☆ ☆

A few days later, Matt and Grandma Emma were back in front of the television set. New York City was throwing the biggest parade Steele had ever seen. It was for the astronaut who had gone on that adventure. For John Glenn. Steele screwed his eyebrows up, looked crookedly at the screen. *I'd like to do something like that,* he thought.

Grandma Emma must have sensed that her grandson was entertaining visions of one day becoming an astronaut. To that end, come birthday or Christmas, Matt Steele usually got a book on space from his grandma. Reading about space flight and watching the televised missions quickly turned Matt Steele into a junkie of space exploration. As he grew up, he marveled at how regular, everyday men would climb into rockets, fly in space—even to the Moon—and then come back home, to be regular men once again. People you could talk to at the doctor's office or at church; guys whose wives shared their blueberry pie recipes with regular ladies who didn't fly in space.

Matt Steele watched the flight updates as if *his* dad was in space. He noted and recorded important statistics and facts from the missions—orbits completed, duration of flight, age of astronaut, speed of craft, astronaut's hobbies, splashdown distance from recovery carrier, and so forth. Question him about any detail of any flight, and Steele had a response.

Question: "What occurred during Gordon Cooper's 19th orbit?"

Answer: "He lost power on his automatic control systems."

By age 7, Matt Steele was seeking a way to be just like John Glenn. Model rockets satisfied that desire. *Things that go up and down are so cool!* he thought.

Before he purchased his first prepackaged Estes rocket, Steele and a couple of his friends decided to test fate by building their own rocket from scratch. They used an empty carbon dioxide cartridge stabilized with aluminum fins and ignited with matches. They began the delicate manufacturing process and seemed well on their way to creating a device that could blow off all of their fingers. Fortunately, one day, while Steele and his pals were riding home on the school bus, a classmate pulled out a copy of an Estes Model Rocket catalog. Steele studied the products and, in the nick of time, concluded that Vern Estes's rockets were a lot safer to use than his own.

Then, scrimping and saving every penny of his meager allowance, he gave his mother the money to purchase his first basic model rocket. Then he waited. Each day, he walked across his parents' farm to the mailbox, to wait for the postman. He sat. He waited, his hands moist with perspiration, his heart pounding heavily in his chest. Eventually, the mailman came and brought a few bills, some junk mail. No Estes model rocket.

This agonizing waiting went on for at least a thousand years, Steele thought.

Finally, one morning his priceless package arrived.

I just can't wait to launch this thing!

That afternoon, mom, dad, and Matt went to the middle of their wide open fields and watched and smiled as that tiny rocket zipped so high into the sky that it nearly disappeared from sight. Then, as quickly as it had raced toward the clouds, it was plunging earthward, right at their heads! Then it was drifting, floating, and bobbing away from them, forcing them to sprint after the tiny parachute and rocket. "Grab it, dad, grab it! We can't lose my rocket."

Seeing the rocket grow larger and larger, the chute billowing in the breeze, forever hooked Matt Steele on model rocketeering. The up-and-downness of the flight became addictive, as if he were the passenger in the nose cone. More flights followed, then more and more until mom and dad thought their son was just a bit obsessed with this hobby. Matt Steele *was* obsessed with rocketeering, but it was no mere hobby: In high school, he began to compete in regional and national model rocket competitions.

Then came the plan: *I'll join the Air Force, become a pilot, then train to be an astronaut.* Steele's life did not follow this blueprint exactly. He joined the Army ROTC and worked on missiles, but instead of pursuing a career as a pilot, Steele accepted an academic scholarship to Kent State University, forsaking his dream of becoming an astronaut, for a more conventional, more obtainable position. He graduated from college with a degree in physics, and carried his knowledge and understanding of missiles and trajectories to military defense contractors Hercules Aerospace and Thiokol Corp.

But there was something missing from these jobs: the *toys.*

Working on real missiles systems was so *adult, serious.* That creative, imaginative part of himself that he had put on hold was prying at his soul. Matt Steele wanted—no, *needed*—to be back in the model rocket business.

In 1984, he and a friend, Chris Pearson, opened North Coast Rocketry, ostensibly a competitor of Estes. In reality, North Coast was little more than a garage shop operation and a means for Steele to buy the tubes to support his rocketeering habit. For eight years, Steele operated the company on a part-time basis, but in 1992, when the defense industry took a downturn, Steele left his job and concentrated on his rocketry company full time. In 1995, North Coast Rocketry officially became a subsidiary of Estes Ind.

Matt Steele has always been a dreamer, and although his dreams and plans have shifted over the years, he will always be that little boy chasing his first model rocket through his family's farmyards. It is as impossible for Steele to grow up as it was for Dorothy to be anything but innocent in Oz. His life is a 24-hour-a-day microworld of model rockets, broad smiles, and dreams being fulfilled.

If you flipped through a family scrapbook, you'd find the awards captured by Steele, along with the certificates garnered and his many accomplishments in the field of model rocketry: Five-time National Champion; five Gold Medals won in international competitions. In the 1980 World Championships, Steele was a member of two Gold Medal winning teams, and he won a pair of Bronze Medals in individual competition. Twice, he stood on the winner's podium as the Star Spangled Banner echoed deep into his soul.

But medals and awards are just a part of the spoils of the business. Then, there are the Born Again rocketeers. It would not be farfetched to say that while some families have family time around the dinner table, the Steeles have it around a model rocket launcher. First, there's mom, Robin, who during the fledgling days of North Coast did much of the company's customer service work. Then son, Cody, 10; Caroline, 7; Kaitlin, 5; and Cassidy, 4. Each owns a rocket or two, and God forgive dad if he embarks on a solitary launch.

Once those tiny missiles leap off the launcher, it's virtually impossible to distinguish the one belonging to the youngest from the one belonging to the oldest child. They all soar so straight, so majestically, you'd think even Cassidy was a veteran of the hobby.

Technically, mom, dad, and the kids have comparable rocketeering skills, but judging by the ever-present joyful smiles, one can only wonder whether it is the children or the adults who are truly the young at heart.

☆ ☆ ☆

So why fly model rockets?

Mark Bundick, President of the National Association of Rocketry, the world's oldest and largest sport rocketry association, has been asked the question countless times.

As a young adult, Bundick found it difficult to explain why he continued to fly model rockets. One day, in the early 1970s, he determined the reason during a conversation with the father of a fellow NAR club member.

In the 1950s, this man had participated in indoor model airplane competitions held in blimp hangars, large auditoriums, and gymnasiums. This, to Bundick, was tantamount to watching a soap bubble, 25 inches across and held aloft by a rubber band-powered motor, float across a large room. The pinnacle of performance was to exceed 30 minutes of duration in a single indoor flight.

Somehow, this man had accomplished the feat, breaking the 30-minute barrier, and shattering the mark set by veteran modeler Pete Andrews. Those who witnessed the historic event cheered wildly, as if they had just seen someone run a three-minute mile. Amidst the pandemonium, a few of the fans hoisted their new hero onto their shoulders, as if he were the MVP of a Super Bowl winning team. They paraded the ecstatic champion around the room, until he came to a rest at a spot just before Andrews, who was also on hand to watch the event.

Andrews shoved his way through the throng to greet the man who had just erased his name from the record books.

"Congratulations, young man," said Andrews, shaking the hand of the new champ. "You're now one of the best in the world at something that nobody gives a damn about and never will."

☆ ☆ ☆

When Bundick met up with this man 20 years later, he asked: "What did you do after that?"

"Well, I kept flying airplanes," the man answered.

"How come?" pressed Bundick.

For the man, the flight of a model airplane constituted a life of its own, separate from this world. First, you designed the airplane, then built it to specifications. Then, you brought the finished model to a competition, confident that it would work perfectly, would fly perfectly. You got someone to time the flight; you set it up, you launched the model.

Then, you walked off the hangar floor, went to the bathroom, chatted with a friend or two. You came back, had a soda, saw another buddy, had another conversation. Twenty minutes later, when you walked back onto the hangar floor, the airplane was still flying.

"That's neat!" the man said, smiling.

Today, whenever model rocket enthusiasts ask him why he still flies rockets, Bundick replies: "It's neat."

☆ ☆ ☆

Mark Bundick was in sixth grade the day his teacher, Mrs. Messick, walked over to his desk. It was never a good sign when your teachers visited your tiny sanctuary; usually, they planned to ask you some impossible question such as "How many sides does a trapezoid have?" Bundick looked down into his shirt; he'd have crawled inside if it were physically possible. But instead of embarrassing Bundick, the teacher dropped a package onto his desk.

Bundick raised his eyes from his chin and peered at the catalog before him. Estes Industries Product Catalog. Bundick had heard of Estes, and, of course, model rockets. The year before, 1964, the national championships had been held in Bundick's home town, Wallops Island, Virginia. Now, the word around school was that the company was trying to persuade teachers to use the model rockets in the classroom.

"You know more about this than I do," said Mrs. Messick. "Maybe you'll be interested in this."

Oh, boy, this is great! Let's see what's in here, thought Bundick as he leafed through the pages.

Mark Bundick *did* know about rockets, as did many of the residents of Wallops Island, site of NASA's Wallops Flight Facility. For years, the facility had been used to launch sounding (special measurement) rockets, and test and research rockets, including the Little Joe, predecessor to the Mercury Redstone. Emmett Bundick, Mark's father, was a member of the range control center team, and he had been working at Wallops since the National Advisory Committee for Aeronautics (predecessor to NASA) had been established in 1945.

Hardly a week passed when NASA wasn't launching a sounding rocket, and the Bundicks had long since developed a routine for launch day. Five minutes before the scheduled liftoff, Emmett Bundick would call home.

"Mark, we're at T-Minus five minutes."

"Okay, dad, I'm out the door."

Mark Bundick would then race out into the yard and wait the prescribed amount of time. Then, BOOM! He could see the rocket zipping overhead; then he could feel the furious roar of the blastoff as it shook the ground.

By age 6, Mark Bundick considered himself an expert on launches and missiles, and from the front porch of his parents' home, he had the inside track on NASA's daily activities. Occasionally, some of the launches involved the release of chemicals into the upper atmosphere. Engineers would fill the sounding rocket with barium or sodium, launch the rocket 200 miles into the atmosphere, then explode the payload into the ionosphere. As the charged particles were acted upon, the mass would change shape and break apart. These things could be seen from hundreds of miles away and, inevitably, the newspapers would soon be flooded with reports of UFOs.

Whenever this occurred, Bundick chuckled to himself: *The general public has no clue what's going on.*

Although Bundick considered himself the foremost juvenile authority on missiles, that didn't mean the local school administration agreed. In the weeks preceding Alan Shepard's 1961 suborbital flight, teachers at Bloxom Elementary School debated whether the students were mature enough to understand the significance of Shepard's mission. If they were not, they should not be allowed to watch the flight in school, some teachers concluded.

That really annoyed Bundick. *What do you mean the faculty thinks I'm not capable of understanding what's going to happen and not appreciate what's going to happen!*

Fortunately for Bundick, reason prevailed, and he and his classmates were permitted to watch Shepard fly aboard a capsule whose predecessors had been tested in Bundick's backyard.

A year later, Bundick was at school recess while John Glenn and *Friendship 7* circled the Earth 116 miles above his head.

There's a guy up there, and he just captured my imagination, he thought.

After that, Bundick watched every Mercury mission on television, always on CBS, always Walter Cronkite. He voraciously read *Life* and *National Geographic,* enchanted by the space pictures that were so much larger than life.

What captured his imagination most was the hardware of space flight: What did the rocket look like? How high would it fly? Would the heat shield work? Would the parachutes open? There was so much to consider.

Bundick's neighbors, including Kathleen Messick, learned quickly that he was interested in anything remotely related to space exploration.

So, with good reason, Mrs. Messick decided to give Mark Bundick the Estes catalog. He flipped madly through the pages, hardly able to keep his attention on one of those wild looking rockets before turning to the next one. *This one is cool! No, this one is better! No, this one is the best!* Wild images of rockets zipping through the sky filled Bundick's imagination. All he wanted to do was to fly one now, right now!

His eyes transfixed by the pictures before him, Mark Bundick reached into his pants pocket and fingered a couple of wrinkled dollar bills. He had been doing odd jobs, and the knot of money was burning a hole in his pocket.

After school, Bundick made a beeline for home. He raced upstairs—Bang!—went the bedroom door. He pulled out a mail order form, sat back, and read and read, and dreamed and dreamed. When his eyelids were droopy and his mind could no longer debate over the relative pros and cons of the rockets, he filled out the order form. The next day, Bundick dropped the envelope into the mail.

A few days later, the package arrived. Bundick tore it open and pulled out the instructions. Like Wernher von Braun studying the schematics of the Redstone booster, Bundick slowly, meticulously, studied each line of the manual. His was a small rocket, though the catalog touted it as possessing tremendous performance capabilities.

Finally, launch day came, and the excitement that had built up in him over the last few days was released, along with the exhaust of the rocket.

Mark Bundick launched that humble model rocket again and again, until, one day, it disappeared from sight, like an old friend who just stops visiting after a while. Bundick was hooked on the hobby, even though his mother tried to convince him that model rocketry was just a fad, unlike the real rockets that seemed to rise with the sun.

But you can't go up to a real rocket, thought Mark. *You can't get close to it, can't fondle it, and you can't be involved with the design of it.*

That's what separated a model rocket from a sounding rocket or a Little Joe: You could build your own model; you got to launch it, watch it rise toward the clouds, watch it nose-dive into the dirt. You yearned to fly it again and again, better and better, make it a perfect mission. At night, when the stars created a milky white luminescence in your bedroom, you thought about taking this hobby to another level. *How far can I take this?*

Mark Bundick took his model rockets all the way to national championships. During one competition, all the conditions were perfect for a textbook flight: a light breeze and a bright blue sky. Bundick was competing in a duration event, flying a rocket-powered glider. The model lifted off like a rocket, then glided slowly, gently back to the ground. The goal was to keep your glider airborne longer than your competitors'.

The launch went perfectly, and the rocket soared into the huge blue blanket. As Bundick walked across the broad field, his rocket continued to soar skyward . . . and soar and soar. Spectators squinted up into the sky, then back at Bundick. Bundick read doubt in their expressions. The rocket continued to race upward until it was a speck against the blue.

Then it was gone.

Bundick walked back to the launch site and was greeted by his colleagues: "Aren't you upset that you lost your rocket?" they asked.

Bundick said nothing, just smiled. He was not the least bit irritated. *Everything was so perfect*, he thought. *I designed and built this rocket and it flew like it was supposed to . . . and it went off into the wild blue yonder.*

☆ ☆ ☆

Vern Estes had been here before. He had witnessed Neil and Buzz and Mike leave their world behind aboard the Saturn V. Estes had been to Cape Canaveral to watch other Shuttle liftoffs, but today's launch was unique. A protegé—a friend—was on board. The date was April, 5, 1991.

Along with Estes, rocketeer Matt Steele and 40 of his colleagues were also there to watch their friend fly into space aboard *Atlantis,* mission STS-37.

Mark Bundick once considered the launches of sounding rockets spectacular; now he found them to be mere popgun flights compared to a Space Shuttle liftoff. Today's Shuttle launch was even more personal for Bundick: His colleague and friend was on board.

These three friends had joined dozens of other adults and children of divergent backgrounds, philosophies, and religious beliefs, bound by a common hobby and the mutual love of an uncommon man, Jay Apt.

Astronaut Jay Apt once told Mark Bundick: "You need to share with others your experiences, no matter how great or small." For that reason, the President of the National Association of Rocketry accepted most lecture invitations, to share with others the joy he derived from rocketeering.

Now, Bundick was sharing his time, his emotions, his love, for a friend and a fellow colleague who had also served on the NAR's Board of Trustees.

As the liftoff of STS-37 drew nearer, Vern Estes's excitement began melding with the familiar pre-flight anxiety experienced by friends and relatives of other astronauts.

He had known Apt for some 28 years, had met him when Apt was a mere kid, just cutting his teeth in model rocketry. *Will everything go perfectly?* Estes wondered now. This was not the inconsequential flight of a model rocket, after all. Next to Estes was his wife, Gleda; she had stood by her husband from the fledgling days of the company, had packed up the rockets and kits and filled the many mail orders that had helped launch the company.

Together, Mark Bundick, Matt Steele, and Vern Estes trained their sights on a rocket bigger and more powerful than any *Astron Mark* or *Astron Scout* they had ever flown.

What was Jay feeling? they wondered, just as a 10-year-old boy wondered how a cricket felt inside a model rocket payload bay.

Emotionally, a small part of each man was about to join Apt, as was a symbol of their collective past, stowed in the spacecraft's airtight storage compartment.

Neither a model rocket, nor a space man, nor a huge Redstone missile had sold Jay Apt on space exploration: a huge inflatable communications satellite named Echo did. Launched on August 12, 1960, Echo 1 was 100 feet in diameter and had an aluminum-coated mylar material that reflected radio signals. Apt was eight years old when *Sputnik I* flew over his hometown of Pittsburgh, Pennsylvania, but the Soviet Union's first steps into space did not rouse his interest. The day he spotted Echo, a tiny round spot drifting at the top of the sky, was the day he began to take notice of space exploration.

Time and again he and his dad drove to the airport and waited anxiously for that shiny globe to pass above them. *This is an outpost of humanity soaring above my head,* Apt mused. Echo not only sparked his interest in space exploration, it piqued his curiosity in the new space program, Project Mercury. Over the next three years, few students at Pittsburgh's Shady Side Academy would follow the exploits of NASA as fervently and closely as Apt.

On the day Alan Shepard bought America a ticket to the stars, 11-year-old Jay Apt sat in study hall, his school books closed. How could a kid study today? An ill-fitting earpiece—the kind that always gave you an earache if you left it in too long—connected him to a transistor radio and the broadcast of the launch. As the other students studied the history of the Soviet Union, or mathematical logarithms, Apt left Pittsburgh some 116 miles below.

Unfortunately, not everyone in his school considered Alan Shepard's flight to be historic. Included in that group of skeptics was the study hall monitor.

"Mr. Apt, what are you doing?" the instructor queried, as Apt struggled to focus on the newscaster's analysis of Shepard's flight.

"I'm listening to the mission," replied Apt, as if listening to the radio were a requisite of study hall.

"No, you are not. Now march right down to the headmaster's office."

This was completely unfair, Apt decided. Something had to be done. How could these people not appreciate the significance of this mission?

Fortunately, Apt didn't have to lobby hard to disabuse the school administrators of their shortsightedness. Two months later, he and his classmates watched Gus Grissom's *Liberty 7* flight on a television set wheeled to the front of the study hall. And by John Glenn's flight, TV sets had been rolled into every classroom. Shady Side Academy had finally caught the spirit of Project Mercury.

Jay Apt followed every minute of every Mercury mission, transforming the radio reports into vivid images in his mind's eye. Television was great—it gave you real life pictures—but when Apt plugged in his transistor radio earplug, he was magically transported to Chris Kraft's flight console. To hear the astronauts' air-to-ground transmissions with the flight control team sparked his imagination and made Apt feel as if *he* were flipping the capsule's switches, monitoring meter readouts. *I want to absorb everything I can about this flight.*

Like many space flight enthusiasts, Apt collected any and all related memorabilia. From the back of a comic book, he ordered space decals that, when applied, converted a household box into a mock spacecraft. But nothing could top the 33 rpm recordings of the transcripts of Glenn's and Wally Schirra's flights. Thirty-five years later, he would still own those rare records.

☆ ☆ ☆

At his 13th birthday party, Jay Apt received a hand-stitched 1963 Estes product catalog from his friend, Joel Davis, who was 15. Apt ordered the *Astron Mark,* one of the first models manufactured by Vern Estes. He was attending an overnight summer camp at Cape Cod the day the rocket arrived at his Pittsburgh home. "Send it to me quickly," he begged his mother and father.

As soon as he received the *Mark,* Apt went to work building it. He awkwardly sanded down the fins. Lo and behold, the fins fit the rocket perfectly, affording the spacecraft a perfect center of gravity. Practice had not yet made perfect, but all in all, the model came together rather well. Apt raced off to one of the camp's baseball fields that was surrounded by a huge fortress of pine trees. *ZOOOOM* raced the rocket into the sky, a streamer trailing in its wake. Apt watched it do its own version of a roll, pitch, and yaw; then, it began its descent.

Uh-oh. It's heading for the trees.

The rocket continued to drift as Apt watched, mesmerized. Moments later, it was gone, lost somewhere in the sky-touching pines.

Jay Apt might have been crestfallen. Technically, he had failed to complete the mission cycle: to launch a model rocket and return it safely to Earth. Yet, he was not the least bit discouraged. *This is a window into the space age. Just like model airplanes must have been for my parents,* he thought.

This modest flight so stirred and impressed Apt, that when he returned to school in the fall, he approached his science teacher, Bill Sayles, with a proposal. "We must start a model rocket club."

And so the Steele City Section of the NAR was founded. For five years, Apt and his colleagues met every other Sunday at a Shady Side field and defied the harsh wind, snow, rain, and whatever else Pittsburgh had to throw in their faces.

In 1964, Apt and his father attended the NAR's national championships held in Wallops Island, Virginia, home of Mark Bundick. A year later, Apt, a skilled rocketeer, captured a National Championship with Dave Baird; two years later, Apt made it two titles in three years, garnering a championship with Jim Davis.

For Apt, model rocketry was the perfect pastime. As a child and young adult, he wore glasses, and he was not from a big, physical family. He enjoyed science and the outdoors, and what better way to combine the two? Model rocketry would be the first steppingstone to a career as a scientist, then as an astronaut, and, finally, as the director of one of the nation's premier natural history museums.

I have a sense that I should be doing the most interesting thing I have the capacity to do, he said to himself.

Apt's road to Cape Canaveral was inevitable, but not linear. He graduated magna cum laude from Harvard in 1971, then completed his doctorate in physics from MIT five years later. From 1976 to 1980 he took advantage of his interest in astronomy by serving as a staff member of the Center for Earth and Planetary Physics at Harvard University. There, Apt supported NASA's Pioneer Venus mission by making temperature maps of the distant planet from Mt. Hopkins Observatory. In 1980, he joined NASA's Space Sciences Division of the Jet Propulsion Labs, and five years later was selected to the astronaut team.

Six years after his astronaut class had been selected, Jay Apt lay inside the cockpit of *Atlantis,* awaiting those first trembles of the spacecraft's engines. Like his many predecessors, he thought not of his friends and family down below, but of the mission ahead of him. He and crew member Jerry Ross—also a longtime rocketeer—were expected to take the first scheduled EVA in five-and-a half years. After gazing wistfully through a telescope at the stars so far above his backyard, Jay Apt soon would be floating among them.

There were so many thrilling aspects to this mission: the research, the ability to peer down at a planet that had so fascinated him for years. Then there was that special package that accompanied him on this voyage.

☆ ☆ ☆

Weeks earlier, Apt had contacted Estes with an idea: "Let's try to take a model rocket into space." A great idea, agreed Estes. What better way to punctuate the impact model rocketry had had on Apt than to carry into space an *Astron Mark or Astron Scout.*

Estes went to work, drawing components from old kits in his private stock, to produce a meticulously crafted *Astron Scout* and two matched *Astron Marks.*

But as Estes waited anxiously at the Cape, he didn't think about the two *Astron* model rockets on board *Atlantis*. He pondered the fate of his friend. He recalled the night before, when more than a dozen of the rookie astronaut's friends, family members, and compatriots had gathered for a pre-flight launching party, minus the guest of honor, who was busy preparing for his flight. The mood had been decidedly upbeat, for Jay Apt was realizing one of his life's ambitions.

☆ ☆ ☆

As *Atlantis* braced to tear free from its hold-down clamps, Mark Bundick also recalled the previous evening's reception. All those people had attended not because Apt was a master rocketeer, or a first-time astronaut, but because he was a friend.

During the celebration, Bundick had stood in a refreshments line, waiting to get a cup of punch. To pass the time, he struck up a conversation with a man in front of him.

"What's your connection here?"

"Oh, I live in the same development as Jay Apt does."

"Oh, really?"

"Yeah, we had a block party and Jay invited us all along."

"That's cool."

Everyone seemed connected.

Finally, the waiting was over, and dozens of people who, like Apt, had experienced the magic of a model rocket launch, were about to witness the mightiest blastoff of their lives. First, the Shuttle trembled, then it shook. As if it had been seized by a giant hand, it leaped into the sky.

When *Atlantis* had disappeared from view, and her world had become strangely silent, Gleda Estes panned the knot of people before her. She spotted Matt Steele and looked into his eyes. They were filled with excitement, they were filled with tears. Gleda Estes said a prayer for her friend. She did not know that at that moment Jay Apt was experiencing the biggest high one could imagine.

☆ ☆ ☆

Mark Bundick has seen some of the 4,000 photographs of the Earth Jay Apt took during his STS-37 mission. Six months after the flight, the two friends met and reminisced about the scientist's rookie voyage into space.

"You don't want to go to sleep," said Apt. "You want to look out the window, and look and look. I couldn't sleep because the Earth was so beautiful, I was drawn to it like a magnet. You can always sleep when you get home."

Apt was fortunate. He had flown with colleagues who were comfortable enough to admit their sense of joy, and confident enough in their abilities to do so.

Jay Apt would fly three more times before retiring from NASA in 1997 to take over as the Director of Pittsburgh's Carnegie Museum of Natural History. During his final three missions, he did not tire of the view of his home planet; nor did he forget the significance of the original Mercury missions.

On the nights before his second, third, and fourth flights, Apt went to the Air Force museum at the Cape. He stood at the pad where Alan Shepard and *Freedom 7* had been launched. Jay Apt had come full circle, back to study hall, with a transistor radio plugged into his ear. Listening, absorbing, wondering . . .

☆ ☆ ☆

On January 18,1998, Dayna Steele received the following e-mail on her Space Store web site:

> Dear Salesperson in Charge,
>
> Heeeeeeeeeeelp! I need your assistance!
>
> I run a 4th/5th grade Young Astronauts Program in (Falls Church) Va. and I use a Plywood Space Shuttle cockpit that is 10 feet long, 7 feet wide, and 7 feet tall, containing 150 toggle switches, some lights, a computer. It seats a flight crew of four, has an intercom system, and a security camera. It communicates with a 16 position IBM computer Firing Room/Mission Control Center. I program the software that is based on the checklists that the astronauts use and we simulate shuttle launch/orbit insertion and reentry/landings. I believe in realism as much as possible and the hardest and still not accomplishable thing to do is to find shuttle space helmets. Unless you have $2,000, NASA will build one for you, or you have to get them on loan, but it is hard to believe that a plastic model is not sold anywhere within reasonable price ranges or that they exist at all. I would think they would be popular, especially with the multitude of space clubs around the country. I just purchased what I thought would be space helmets from a toy company but they look like light bulbs. I might be able to modify them to look reasonably like shuttle helmets but it will be one heck of a challenge. Where can I get reasonably priced $30 plastic shuttle space helmets? Do they exist?
>
> Thanks,
>
> Mike Stanton, Belvedere Elementary
> Young Astronaut Chapter Leader

"I have a helmet!" Mike Stanton announced triumphantly. "It's on loan from NASA. We took photos and measurements and now we will re-draw the diagram with measurements, so if someone can do so . . ."

If is a word with which Stanton was painfully familiar. For nearly a year, he had scoured the country, searching in vain for an affordable replica space helmet for the students in his astronaut club. Queries to politicians yielded sympathetic encouragement, but scant concrete support. Toy manufacturers considered the potential market for a plastic astronaut's helmet, then declined. *Sorry, we don't have the means to make such a space helmet.*

But now, six months after Stanton's desperate Internet E-mail had introduced his Young Astronauts Program to a dozen space enthusiasts, the 42-year-old engineer was one small step closer to creating complete *realism.* Before him sat a legitimate NASA launch and reentry helmet, complete with visor and communications hookups. While a single unit would not satisfy the needs of the four-person flight crew, Stanton hoped the helmet could be used as a prototype. Maybe with these exact dimensions and schematics, some enterprising director of marketing at some niche toy company might say: "Yes, we feel there is a market out there for this product. We'll manufacture it."

A generation earlier, Scott Pelley had discovered a whole new world through a four-by-six-foot box, a handful of space decals, a felt-tip marker, his imagination, and a sky full of stars. But as the technology of space flight became more sophisticated over the years, space enthusiasts began to demand equally sophisticated toys with which they could launch their own pretend missions.

To wit: Stanton's troop of 10- and 11-year-old steely-eyed missile kids were accustomed to using real computer scripts to run their mock Space Shuttle missions. They were accomplished at solving the programmed casualties that stalled the launch or de-orbit sequence of events.

YAP's flight crew wore orange flight suits, complete with NASA patches, and they flew their missions in a cockpit constructed from balsa wood, adorned with Plexiglas windows that were shaped like those of a real Shuttle. A black tarpaulin, suspended on a pegboard, surrounded the shuttle, and created the illusion of outer space. Inside the craft, the YAP shuttle commander and pilot flipped the toggle switches in accordance with the flight checklist. Headsets and an intercom system linked the astronauts to a Capcom back at the Firing Room/Mission Control Center.

There, flight controllers, designated FLIGHT, TELMU, EECOM, and so on, ensured the mission went according to plan. These 16 systems engineers bore the same Gene Kranzian rock-hard, impassive countenance. But unlike the former NASA flight director with the hallmark crew cut and the stern demeanor, one YAP flight engineer possessed a Kewpie doll face and wore her coal black hair in a loose pony tail. Another blond-haired boy had inquisitive eyes that never seemed to blink.

All in all, this was as close as any 10-year-old space nut could get to flying in space. Except there were no helmets.

Stanton recognized that this one missing accouterment would not prevent his students from learning basic Space Shuttle operations. A space helmet would not make a YAP pilot respond more efficiently under pressure. No protective helmet would afford an EECOM a more thorough understanding of the math and science that were a major part of the program.

But Stanton wanted the helmets for himself. He felt an incompleteness without them. As Mark Bundick, Matt Steele, and so many other space enthusiasts flew model rockets because they replicated the flight of a real rocket, Mike Stanton needed those helmets because it would bring his program one step closer to the *real thing*. Moreover, for a longtime enthusiast of space exploration, this was as close as Stanton would ever come to firing a Shuttle's thrusters.

Fortunately, Stanton had much to do. He was preparing to create a universe for his students, a homemade galaxy that would convince any 10-year-old she were 116 miles in space. First, he would punch holes in the bottoms of empty shoe boxes. Then, he would fill the bottoms of the boxes with sparkling foil. Through the foil, he would string small Christmas bulbs, then replace the lid with a piece of black paper onto which he would mount the star constellations. This mock universe would be attached to each shuttle window, and when the lights were illuminated, the YAP astronaut crew members would feel as if they were circling the stars.

A chill rippled through Stanton as he pondered this project. *I just look at it as God saying "Stay on Earth and teach kids about the space program."*

☆ ☆ ☆

Although it took Mike Stanton 34 years to recognize his lot in life, his fate was sealed the day he was born.

Routine post-partum examinations had revealed that Stanton's head was slighter larger than normal. "Keep an eye on his head, Mr. and Mrs. Stanton," the attending pediatrician warned. "It's bigger than normal, and there could be something wrong here."

The physician's concern was warranted. As the weeks and months passed, Mike Stanton's head grew larger and larger as unrelieved fluid caused his brain to swell. As Stanton approached his first birthday, doctors determined an operation was necessary to stem the swelling and drain the fluid around the brain.

Doctors peeled away the skin on either side of Stanton's skull, then inserted tubes to drain the amassed fluid. Finally, to prevent a reoccurrence later in the child's life, the surgeon inserted pins in Stanton's skull. While the operation likely saved Stanton's life, it came with a price: The pins that prevented the buildup of brain fluid could never be jostled or moved.

Stanton would fully appreciate the significance of those screws years later when he enlisted in the Coast Guard and then the Navy: "No flight training or deep-sea submarine work," they told him. "You could suffer decompression and that might affect your brain."

Mike Stanton was grounded, just like Deke Slayton, just like Al Shepard.

☆ ☆ ☆

Some children are born to be John Glenn, dreamers who fantasize of flying off to distant galaxies. Others, such as Mike Stanton, see themselves as the next Wernher von Braun, designing the rockets of the future. Stanton was 6 years old the day he approached his mother with a small model covered wagon has father had built.

"Why is the forward wheel of the wagon always smaller?" he asked his mom.

Marcie Stanton rolled her eyes. Like her husband and their other three children, she had little interest in space exploration and the attendant mathematical and scientific conundrums that intrigued her son until all hours of the night.

"I don't know, you tell me," she responded.

"The back wheel is larger, and when I turned it, the front wheel was spinning faster."

Marcie Stanton went blank. Her son wrinkled his face in dismay.

"I'm worried about the circumference of the wheel and how it gauged the distance."

What is he talking about? Marcie Stanton asked herself.

Unable to obtain an answer to his questions, Mike Stanton withdrew to his bedroom to ruminate over the problem of the back wheel.

☆ ☆ ☆

If solitude is the prize for ingenuity, then Mike Stanton was the loneliest kid in Palmdale, California. As he grew up, his interest in space flight blossomed. He watched all the NASA missions, fascinated with the rockets and the technology. At age 8, he received a model Gemini capsule from his father, Harry; now Stanton could play NASA engineer, using his Gemini capsule to plot trajectories of pretend flights. *How did rockets fly? Why did they curve upward instead of rising in a straight line? Where did the astronauts go after the frogmen plucked them out of the water?*

All alone in his private universe, Mike Stanton retreated to his bedroom—his Mission Control—to find the answers to these daunting questions.

He was 11 when he decided to put his knowledge of math, science, and related flight trajectories to work. To mimic the launch of a rocket, Stanton used a few cylindrical sticks of fireworks, each about an inch long. They were of the type commonly used on the Fourth of July: You placed them on the ground, lit the fuse, and up they soared. But Stanton had a better idea: *Let's test the propulsion of these fireworks,* he thought.

Collecting these makeshift mini-rockets, Stanton raced across the street to the schoolyard of Juniper Elementary. He theorized that if he hurled the lit fireworks into the air, when the thrust kicked in, they would leap forward like a Saturn V during second stage ignition. Stanton grasped one of the fireworks between his thumb and index finger. He lit the fuse and hurled the firecracker like a javelin, sending it spinning toward the roof of the school. *It'll burn out before it reaches the school,* he thought. The firework rotated like a tiny satellite, then—*Boom*—it leaped forward and tore right into a gusting wind.

Oh no. The wind. I didn't account for that.

Suddenly, the firework was plummeting out of control, like one of the early Atlas missiles. *It's heading for the neighbor's yard!* Stanton double clutched as he saw the sputtering firecracker land in a knot of tumbleweeds. Next thing he knew, the fuel had ignited the arid plants, transforming his neighbor's yard into a miniature brush fire.

"Fire! Fire!" screeched the panic-stricken Stanton. He made a beeline for home. Stormed up the steps. Yanked open the front door.

"Mom, get the hose out back, there's a fire!"

As his mom sped out the back door, Mike Stanton grabbed the hose from the front yard. In one huge leap, mother and son converged upon the fire, where they were quickly joined by the astonished neighbor, a man named Andy. Andy also was wielding a green garden hose and was doing his best to douse the spreading flames.

Within minutes, the nervous trio had extinguished the fire, narrowly averting a disaster, and eliciting a group sigh of relief.

Later that night, Stanton's parents brought up the incident.

"What happened?" asked his mom.

I'd better come up with a good excuse, or else they might discipline me, thought Stanton.

Delving into his mental file cabinet of technical claptrap, the 11-year-old proceeded to stupefy his parents with a brilliant—albeit convoluted—dissertation on trajectories and the influence of wind. Force of thrust . . . inverse to the propulsion of the rocket . . . versus the direction of the wind . . . in correspondence to torque . . .

No punishment was levied that evening.

As a teenager, Stanton's fascination with mathematics and science encouraged him to be not an astronaut but an inventor. *Maybe I could develop something for space, but what?* he thought.

One day, he decided to build his own pinball machine. *A pinball machine is like a pool table,* he considered. *You shoot the ball against the wall at a 45 degree angle and it will bounce back 45 degrees in the opposite direction.*

Among the game's components were 115 volts AC, a steel ball bearing, and aluminum foil. Stanton figured that when the ball ricocheted off the foil, it would conduct electricity which would, in turn, illuminate a light bulb. Lo and behold, fireworks.

Driven to succeed like a mad junior Einstein, Stanton built the prototype of a game that would surely change the recreation world. Unfortunately, while diagramming the schematics, he neglected to consider electrical resistance. When the tiny steel ball began careening around the board, it nearly blew up the circuitry in the house. It sent a charge through the game's circuitry, producing a ball of smoke and a flame that told Stanton his invention needed more work.

Back to the bedroom.

Left to his own devices, Mike Stanton spent much of his idle high school days hunched over his bedroom desk, working on trigonometry or trajectory problems. *Hhhhmmm, let's draw a circle, then make all the angles of that 360 degrees. Ah-ha! Now, let me plot the X and Y coordinates . . . Hmmmm.* Stanton's natural attention to detail was an aide at his drafting table, where he created three-dimensional drawings of rockets and other architectural designs. To draft was to be the inventor, the explorer of his own fertile imagination. In this introspective world he had a mentor, his Aunt Pat.

Patricia Kalata was a professional draftsperson who had worked at Edwards Air Force Base, and had contributed drawings to the base. If mom and dad were uncertain of their son's interest in space-related subjects, Aunt Pat was 100 percent gung-ho. She had, after all, given her nephew his first drafting set. "Draw, dream, pursue your goals," she encouraged Stanton. And so he did. Stanton quickly and easily became skilled at this hobby. At a county fair, he was awarded a blue ribbon for his three-dimensional drawing of a house. He also drafted a set of electrical schematics, which were accepted and registered by the Air Force. Finally, to cap his high school career, Mike Stanton earned a prestigious award for his drafting and architectural drawing.

At age 18, seeking to further his interest in mathematics, science, and rocketry, Stanton enlisted in the Navy. He went to work on missile systems, his passion. But when Navy doctors reviewed the X-rays of his head and saw the pins, they declared him unfit for flight and submarine work.

Although often relegated to a Navy base or an aircraft carrier, Stanton's work challenged and captivated him. The launch of a surface-to-air missile was a personal mission for Stanton as well. First came the exhilarating launch, then the tracking of the missile's trajectory: The missile soaring and arcing upward, contrails streaming in the broad, boundless sky. The wind catching the projectile. Suddenly, Stanton was back at Juniper Elementary School, his firework veering into the neighbor's yard. *The contrails . . . the smoke, the burning tumbleweed.*

☆ ☆ ☆

Over the next 16 years, Mike Stanton continued tracing Navy missile launches with the same satisfaction and interest he accorded NASA's manned space programs. His job took him from California to Charleston, South Carolina, then to Virginia Beach. By 1990, Stanton was settled, married, and the father of two girls. That year, the door to his imagination was pushed wide open.

Kellie Stanton was 9 years old the day she eternally changed her father's life. She had been attending a public school four days a week, but had been bored and felt unchallenged. One day a week, she commuted to a private school in Virginia Beach called Old Donation Center for the Gifted and Talented. ODC offered Kellie the chance to study among her intellectual peers, to learn sign language, to participate in a unique course called the Young Astronauts Program. YAP was a nine-week curriculum that used the Space Shuttle Program to expose students to math and science, but in a hands-on way. To Kellie Stanton, YAP made school interesting; it made space exploration an obsession. One day, she told her father about the junior space program.

"I know you love watching the Shuttle and stuff, but can you help the teacher teach it?" Kellie asked her dad.

"Sure, why not?" replied her curious father. "What is the Young Astronauts Program?"

On October 10, 1990, Mike Stanton found out. When he arrived at Old Donation, Kellie Stanton's teacher was discussing satellites and digital imagery, explaining how information was traced from the satellites to the radar station.

Listening to this lecture made Stanton feel 15 years old again, sitting at home, working those trigonometry problems to death, figuring the Xs and Ys, the angles. The program intrigued Stanton, just as Project Gemini had enticed him to the world of space exploration. He asked himself, *What can I do to repay them for letting me help teach this chapter? I can make this class as realistic as possible.*

So one day, Stanton brought two walkie-talkies to the class. Afterwards, to the school library he brought the radio set, along with four chairs, the Shuttle Procedures Manual, and the YAP students. Using the walkie-talkies and a Shuttle launch script from the manual, Stanton transformed a roomful of desks and books into Old Donation Space Center, complete with a launch control center and a crude shuttle orbiter, fashioned from the chairs. Then, line by line, the astronaut/students worked through the script, reenacting the basics of a Shuttle launch.

Ninety minutes a week, for six weeks, Mike Stanton became unofficial Director of Flight Operations for his daughter's YAP chapter. For six weeks, he was back in tune with the space program, speaking a native dialect: math and science, space terminology, and nomenclature. *T-Minus ten seconds and counting* . . . This pretend outer space world made Kellie Stanton's eyes sparkle like tiny stars; more important, it connected Mike Stanton to others in a world he had inhabited alone for so many years. When the session concluded on November 21, 1990, Mike Stanton vowed he would not sever this emotional umbilical cord.

For Stanton, the decision to establish his own YAP chapter was as cathartic as NASA doctors telling a grounded Al Shepard he was finally flightworthy. Stanton, now 34, had been successful and content, both at home and at work. But he had also felt incomplete, deep down inside his soul. He had wondered why. Now he knew. YAP allowed him to emotionally remove the skull pins that had grounded him. That urging inside of him, to expand beyond his job with trajectories of missiles and rockets, had been suppressed for so long. Now it had finally found its release valve. Maybe Mike Stanton couldn't physically sit in *Columbia* alongside Bob Crippen, but through the

Young Astronauts Program, the little boy in him could finally fly to the Moon.

It is a dream waiting to happen, he thought. And as that dream formed his reality, Stanton finally rode *Friendship 7* into low Earth orbit.

He had questions, however: *What will the students retain? Will they use the program to better understand their school subjects? Will they be able to correlate these scripts to an actual Shuttle launch and reentry?*

Stanton believed they would, for he had witnessed a child's profound power of understanding, particularly at a subliminal level.

During Mary Stanton's pregnancy with Kellie, she and her husband sang songs, read stories, and chanted the alphabet to their unborn child. The couple believed that, even in the womb, a child learned, sensed, felt.

"A, B, C, D, E, F, G," the couple sang. "H, I, J, K, L, M, N, O, P, Q, R, S, T, U, V, W, X, Y, and Z . . . Now I know my ABCs, next time won't you sing with me?"

After reading dozens of fables, singing countless songs, reciting the alphabet from Z to A, Mary Stanton was ready for the Cesarean section that would bring her child into the world. As his wife underwent surgery, Mike Stanton paced anxiously in another room. He waited. And he waited some more.

After awhile, a nurse whizzed past him, carting a newborn child to the nursery. Minutes later, a second nurse emerged from another room and walked up to him. "Who are you?" she inquired.

"I'm Mike Stanton. I'm waiting for my daughter to be born. My wife's having a Cesarean section."

"Oh, that must have been your child," the nurse blushed, referring to the newborn who had been whisked away minutes earlier. "We're sorry."

Stanton sped off to the nursery, which, he discovered, was brimming with screaming newborns who suffered from varying degrees of hunger, colic, or loneliness. It was like a human assembly line, and the babies all looked and screamed alike. Which child was his? *Where's Kellie?*

Then, a voice spoke inside Stanton's head: *The ABCs . . . Let's put this to the test.*

"A, B, C, D, E, F, G . . . " sang Stanton into the crowd of hollering, toothless, bundles of pink lumps. "H, I, J, K, L . . . "

Suddenly, Stanton heard the din break ever so slightly. Amidst the cacophonous screams, he perceived a muffled sob. Finally, the group cry was one scream weaker. During his recitation, one of the babies had stopped crying. *Could it be . . . ?*

Still in his finest voice, the first-time father snaked his way through the nursery, past the rows of incubators, keeping his focus on that one calm child: the *ABCs* baby. He reached the bassinet, ran his eyes from the baby's toes to its head. Searched for the identification card. *Kellie Renee Stanton.*

It was story time.

Four years later, the Stantons repeated this procedure at the birth of their second daughter, Katie. Same result.

As a result of these experiences, Mike Stanton had little doubt that his YAP students would grasp the curriculum. *Just as long as I make it interesting and realistic,* he knew.

For two weeks in December, 1990, Stanton traveled to the Langley Teachers Resource Center to collect information on space exploration and related topics. He returned home with stacks of research materials, tapes, and detailed information on the Shuttle: how it flew, how it worked. Everything a 10-year-old would need to know about the current manned space program.

In January, 1991, Stanton contacted the Young Astronauts Council in Washington, D.C., to establish his own chapter. The Council oversaw the 25,000 YAP chapters worldwide. He went to work soliciting members for his chapter. He approached a local elementary school, but was turned down. Then, within a two-week period, the Girls Council Inc. of Virginia Beach, Virginia, and Chesterfield Heights Elementary School in Norfolk expressed interest in the program. In early February, the first Girls Council and Chesterfield Heights YAP meetings took place. Mike Stanton and his astronauts were on their way to the stars.

No program would be complete without a curriculum, but Stanton also felt his Young Astronauts Program required more than the Shuttle computer scripts to make it interesting. It needed to replicate NASA, needed a realistic mission control room and a believable replica Space Shuttle. Stanton returned to his drafting table, where he laid down the blueprints for the chapter's first model spacecraft.

Stanton cleared one hurdle when a local lumber yard agreed to donate the lumber for the space ship. Using two-by-fours and plywood, Stanton quickly built the shuttle's foundation and the forward section of the cockpit.

Over the next few months, he honed the program's curriculum while he pieced together the YAP shuttle. The orbiter's frame was expanded and window frames were also installed.

One perfectly quiet summer night, the type of evening when a high Moon beckons, Stanton was outside, applying a few touches to the shuttle. He was sitting on a chair, alone in his little craft, when the feeling struck him.

The stars . . . the black night . . . to fly.

Stanton felt a wave of dreaminess sweep over him, as it did when he watched the early space missions. *I wish I had a camcorder; I could shoot from this cockpit and see the stars, see the rotation of the stars. I could pretend I was in space, just like a kid.*

By mid-1992, the YAP shuttle was as portable and reusable as its NASA counterpart. Stanton disassembled and reassembled the shuttle for use during school visits. Such trips showed curious students that, with the computerized scripts, they could perform their own shuttle launch, orbit, and landing procedures, just like NASA's astronauts. Stanton furthered the connection between NASA and the Young Astronauts Program by inviting Shuttle astronauts Ken Reightler, Jr. and Jeff Wisoff to speak to his class.

But YAP wasn't all math, science, liftoffs, and logic, it was a rocket ride to the Moon, even for children who were too ill to climb out of a hospital bed. In May, 1992, Stanton contacted representatives at Norfolk, Virginia's Ronald McDonald House about doing a brief version of the program at the Children's Hospital of the King's Daughters. Hospital administrators explained to Stanton that while the parents were being briefed on the cancer treatment their children would receive, the children required entertainment.

"We would be pleased to present the entertainment," said Stanton.

On June 15, Stanton visited the children's hospital and made arrangements to put on a week-long Young Astronauts Program. Two days later, he appeared on the hospital's internal television network to explain to the children how YAP worked and how they would be involved. From their hospital beds, children called Stanton on the phone and wrote notes to him: "How do the astronauts get into space? How far is it to the Moon? What can I do to become an astronaut?"

Stanton went to work assembling the shuttle on the fifth deck of the hospital. He carried up each piece by elevator, past the wide-eyed children who had gathered to watch. Stanton was reviewing the electronics of the craft—the switches, the controls, the computer—when one of the children struck up a conversation with him about the cockpit. His eyes were tired but triumphant. "I have had 25 operations," he told Stanton. "But I'm leaving tomorrow."

This could be a very satisfying experience, thought Stanton.

On June 18, Stanton reviewed launch and de-orbit procedures with the children. Typical YAP students would have clamored to get inside the shuttle, but many of these children were burdened with IV bottles that hooked onto portable tripods. To load these children into the craft, along with their tripods, was impossible. Stanton eyed the wondrous children and felt his heart sink.

How do I solve this problem? Stanton wondered. The answer lay in the hardware: Stanton modified his craft so that the tripods could remain outside the craft. The IV bottles were still attached to the children, still dispensing critically-needed medicine to each; but for the moment, they had left behind their debilitating illnesses to become pioneers of a new world.

☆ ☆ ☆

Some YAP missions heal the soul; virtually all steel a student's mind, filling him or her with a deep sense of accomplishment. During one typical mission, Stanton was stationed in the shuttle cockpit. He was focused on the mission pilot, a little girl who had always wanted to be an astronaut.

She's so intrigued, Stanton thought, as he watched her study the mission checklist. It was a challenging script, he knew, and by her furrowed brow and frown of concentration, he could see she was battling to learn the proper sequence of things, to flip the toggles at the right time. As if the shuttle would plummet out of orbit if she erred.

The pilot raced through the checklist, following procedures perfectly. She was gaining confidence, Stanton could see. Her eyes were live, tiny, fiery worlds. Thought Stanton: *My Katie's in her own little world.*

Journalists

Howard Benedict was struggling to relax, when the telephone rang. Earlier in the year,1990, he had retired from the Associated Press, leaving behind a job that had propelled his career as high as any rocket booster had launched a man. For nearly a generation, he had explained in simple, direct terms, the complexities of space flight to a nation of enchanted people who merely knew that if we went higher, farther, faster, we would beat the Russians. Now the longtime Associated Press reporter was satisfied to take a less stressful approach to life and space flight.

But when Benedict recognized Henri Landwirth's voice at the other end of the telephone line, he figured his day of leisure was about to come to a precipitous conclusion.

"Howard, I want to talk to you about the Mercury Seven Foundation," Landwirth said.

Landwirth was one of the founders of the organization, and he was a long-time friend of Benedict's. Back when Cocoa Beach was a hotbed for hotshot pilots, Landwirth had managed a hotel called the Cape Colony Inn, temporary home to both Benedict and the Mercury 7 astronauts.

"How can I help?" replied Benedict.

"We're kind of rudderless now and we need an executive director for the foundation. Why don't you come work for us."

Benedict paused and briefly considered Landwirth's offer. The Mercury Seven Foundation was a worthy venture, he understood. In

1984, the surviving six Mercury astronauts, along with Betty Grissom, widow of Gus Grissom, Landwirth, and William Douglas, the Project Mercury flight surgeon, had collaborated to establish the organization. Benedict felt the Mercury Seven Foundation was more than just a non-profit organization that provided scholarships to upper level college students and students working toward their masters or doctorate degrees in science or engineering. The foundation was, in a sense, the Mercury astronauts' legacy, a timeless source of inspiration for those who grew up dreaming to be astronauts. The scholarship money raised by the foundation helped draw these students one small step closer to a flight in outer space.

"Henri, I don't want to do a full-time job," replied Benedict. "All I want to do is take an early retirement, and do some writing and relaxing."

"Well, think about it," replied Landwirth. Then he hung up, hoping to hear back from his friend.

Three days later, Howard Benedict's phone rang once again. This time, the caller was the president of the Mercury Seven Foundation, the first of the original astronauts to fully trust Benedict as a reporter: Al Shepard.

"We'll just try it for 25 hours a week," said Shepard, trying to sell Benedict on the job. "You can set your own hours. We need someone to bring order out of this chaos. We're losing our focus."

What could Benedict say? This was Alan Shepard, after all. Who could reject the first American to fly in space?

"OK, I'll try it," Benedict said.

Two weeks later, Benedict realized he had assumed a full-time job. So much for relaxing and writing. He also recognized he had a challenge before him. Benedict, who had spent most of his career beating newspaper wire deadlines by T-Minus two seconds and counting, enjoyed a challenge. What could be better than working with his old friends?

☆ ☆ ☆

After eight years at the helm, Benedict has more than refocused the organization. In 1986, the Mercury Seven Foundation raised its first seven $1,000 scholarships. In 1995, the Board of Directors, realizing the surviving astronauts could not raise funds forever, expanded membership to include other astronauts. It also changed the name of the organization to the Astronaut Scholarship Foundation.

Today, the foundation annually awards 15 scholarships, each worth $7,500.

Benedict's office, located in a facility at Kennedy Space Center, (also the home for the U.S. Astronaut Hall of Fame and Space Camp-Florida), is a tribute to the past. Framed pictures of the Mercury astronauts adorn the walls. With a glance at these photos, Benedict slips back in time, from 1990 to 1980 to 1970 to 1961. To an era when Shepard and crew rewrote the science fiction books.

☆ ☆ ☆

Howard Benedict's inevitable journey to the Astronaut Scholarship Foundation began the night an ominous red sun set upon the nation. On October 4, 1957, Benedict was a seasoned reporter with the Associated Press, spinning human interest stories for the wire service's features department. By chance, he was also the only news reporter in the department's New York City office that day.

Benedict was working on an in-depth feature article when his interest was roused by a stunning bulletin that flashed across the news wire. The Soviet Union had just launched *Sputnik I,* the first man-made Earth satellite.

"Geez, we gotta have a feature for Sunday on what this *Sputnik* is," said Benedict's shocked editor.

The puzzled reporter looked at his boss: "What the hell is this satellite?" Benedict asked. "I'm not qualified to write this."

"Well, there's nobody else qualified to write it either," retorted the editor. "Everybody's new at this game."

The reporter in Benedict went to work. *Who do I know who can give me some information on this thing? How about Wernher von Braun? He's been writing those wacky magazine articles about flying to the Moon.*

Benedict telephoned the ex-German rocket scientist who had helped to develop Germany's devastating V-2 vengeance missile. "What is this *Sputnik*? What does it mean to the nation?" Benedict asked. Von Braun shared a wealth of information on ballistic missiles and on the pioneers of rocketry, including Rumania's Hermann Oberth and America's Robert Goddard. Then he boggled Benedict's mind with fanciful visions of man's future in outer space.

Pie in the sky nonsense, thought Benedict afterwards. *What is this* Sputnik *and where could it lead?*

Despite his own skepticism of space exploration, Benedict forged onward with the article. He contacted a few reputed experts at the Pentagon and matched their quotes with the material provided by von Braun. He wrapped up the piece in a few thousand concise—and often unbelievable—words.

That Sunday, newspapers around the world ran Howard Benedict's fantastic article on *Sputnik.* There was nothing esoteric or technical about it, just information and images that seized a reader's imagination: The Soviet Union dropping bombs on the U.S. from outer space; weather forecasting from outer space; space stations; trips to Mars.

Stuff that scared the soul out of people.

With one article, Benedict had become an expert on space exploration. Of course, nearly all of his peers were neophytes on the subject. Thereafter, whenever the wire service required a feature story on a space-related topic, Benedict got the assignment.

In 1958, Benedict commuted more frequently to Cape Canaveral to cover the early satellite and rocket launches. Covering a semi-steady beat struck an emotional chord with the newsman. Years earlier, he had earned his first press badges as a beat writer. Now he was growing weary of the plodding, stale feature writing business; at times he would spend an entire month researching a story. *This is too slow for me,* he thought. He craved to return to spot writing, covering events as they unfolded, not when the events had become old news. Space exploration, Benedict figured, would always be new, fresh, ever-changing. *Suppose we did build a space station? Suppose maybe one day we will go to the Moon? Suppose von Braun is right?*

In 1959, Benedict got the chance to return to spot news, when the AP opened a Florida bureau designated to cover NASA and the space program. Benedict now belonged to a small, elite circle of journalists whose words and pictures would change the way people thought about space flight.

Included in this journalism aristocracy was 24-year-old television/radio newsman Jay Barbree. *This is an exciting place to be*, he thought while watching missile after missile blast off from the Cape. The other thing Barbree noticed, when covering live launches, was that NBC did not have a permanent representative in the area. Eagerly, he offered his services.

On July 21, 1958, Barbree received a call from an NBC executive.

"We received your letter. We'd like to try you out," the editor said. "We understand there's a satellite launch next Tuesday."

"That's correct," said Barbree. "There's a mouse in the nose cone nicknamed 'Wicky Mouse.' It drinks its water from a wick."

The editor broke into peals of laughter. "You're hired."

The Wicky Mouse launch was a financial windfall for the enterprising reporter. He filed three hourly reports for NBC and a world news roundup, and walked away with $155. Not a bad night's pay.

But Barbree would likely have done the job for a dollar and a half, for there was no question in his mind that the nation and NASA were going to go up and out. To ferret the often buried stories, Barbree first had to know the hardware, the rockets and the people involved in space exploration better than any other reporter.

Consequently, on the day the world was introduced to the Mercury 7 astronauts, Jay Barbree was at Wright-Patterson Air Force Base in Ohio, researching a story for the NBC show *Monitor*. He was scheduled to report on the rigorous testing the astronauts had undergone during the selection process. To authenticate the broadcast, Barbree wore a microphone throughout the testing. He spent time in an isolation chamber that was so silent, so still, he could hear his heart beat. He was dipped into a pool of ice. He was whirled in dizzying circles by a centrifuge that subjected him to stomach-churning G forces. His news report was human and honest. Fortunately, his lunch obeyed the laws of gravity and stayed down.

While researching the story, Barbree learned that NASA had selected the seven men to fly Project Mercury. Included in the group of astronauts was Gus Grissom, a native of nearby Enid, Ohio. *Now here's a story,* thought Barbree. He interrupted his schedule, grabbed a camera crew, and flew to Enid to interview Betty Grissom and her sons, Mark and Scott.

Suddenly, Thor rockets and dispensable pilots named Wicky Mouse had been supplanted by fragile human beings. People who not only represented the future of manned space flight, but of humanity itself.

It was natural for the nation to cheer its new heroes. It was difficult for even the most staunchly objective reporters such as Barbree, Benedict, and CBS's Walter Cronkite, to remain cool to the pioneering spirit that was embracing the country. After all, their lives were becoming Project Mercury, space flight, and the prospects of man flying to the Moon.

For Benedict, the NASA beat was a seven-day-a-week, 24-hour-a-day job that would not slow until after the flight of Apollo XI. Between the unmanned rocket launches, the satellite launches, and preparations for the manned Mercury flights, Howard Benedict virtually lived at the Cape. He and the other core members of the press set up temporary shop in a shack not far from the launch site. They sat for countless hours in there, waiting for rockets to either lift off and fly away (to the satisfaction of NASA), or to explode and crumble into the ocean (much to the chagrin of NASA). The newsmen passed the time playing poker, enough poker to make a man rich or poor, if the stakes weren't so low. When everyone had grown weary of full houses and flushes, Benedict picked up an aviation or space magazine and read it cover to cover. He learned early on that there was no such thing as too much knowledge, for when you talked to some of NASA's brightest engineers and technicians, you had to know what you were talking about.

Most of the missile launches occurred at night. In the pre-computer age, NASA engineers used the stars as reference points against which they could check the telemetry of the rocket. The shifts were long, often stretching to dawn, sometimes going into the next day if a countdown was placed on hold. One time, a pair of engine shutdowns forced NASA to twice place on hold the launch of an Atlas rocket. For 36 interminable hours, Benedict sat in the shack, playing cards until spades looked like diamonds. Finally, the troubled launch took place.

Poker, press conferences, launches virtually day after day. Endless deadlines. It was a hectic life, a tiring life, an exciting life for Benedict, but one that allowed for few social activities. Fortunately, Benedict's wife Joy shared his interest in space exploration. Joy Benedict was a writer for *Florida Today* who frequently wrote about the women of NASA, particularly the astronauts' wives. She was also a swift typist and a whiz with a teletype machine. That was a boon to her husband, who preferred dictating his stories to be typed.

A Howard Benedict article was fair, balanced, and never subjective. Opinion was best left for the columnists, he concluded. Fair meant telling the story accurately, at any cost. Benedict, Barbree, and Cronkite each covered the warts of the program as closely as they covered the successes. When a Redstone or Atlas exploded seconds after lifting off, they reported the miscue in painstaking detail. They were not reluctant to criticize or question a NASA decision, even if that rendered some NASA administrators ruddy-faced and white-lipped with anger.

Cronkite also recognized that a reporter should be objective and should write the truth, devoid of personal interest. Yet, he also believed that for a writer to not be excited or intrigued by the prospects of traveling into outer space, he would have to be brain dead.

In fact, Benedict felt no guilt about rooting for the astronauts, as long as it did not compromise the integrity or objectivity of his work. *Newspeople aren't supposed to be cheerleaders for the program they cover,* he considered, *but we have become cheerleaders for the astronauts and rightfully so.*

Thus was the challenge for Benedict, Barbree, and Cronkite: to separate the objective reporter from the thrilled human being.

Painting an objective yet compelling portrait of space flight was relatively easy for all three. Yet, researching their stories was significantly more challenging, since the astronauts and their families had signed an exclusive contract with *Life* to publish their life stories. While the agreement limited the astronauts' involvement with the ever-probing media, it severely hampered the other journalists. Benedict and his peers were forced to settle for the astronauts' and the agency's bland, regurgitated comments during press conferences. Even before the first

American had flown in space, the all-American smiles of NASA's astronauts had graced the covers of one of America's most popular and respected publications. But, innocuous comment by comment, the astronauts revealed more of themselves to the press and to America. Before long, these often reticent public figures befriended some of the reporters, in particular Benedict and Barbree. But first, an American would have to put his life and the morale of a nation on the line.

The reporters were particularly anxious when it came time for the first American to fly in space. *Are we rushing things?* Cronkite wondered. Images of Redstones and Atlases exploding in the Florida sky weighed heavily on his mind. *Are we pushing the envelope by putting a man on board at this stage?*

The man NASA chose for that first flight was Al Shepard. Publicly, NASA walked the politically correct line, asserting that all seven men were immensely talented, all capable of taking America's maiden voyage into space. But privately, the selection committee told Barbree that Shepard had garnered the initial mission because he was the "smartest of the bunch. If something had to be taken care of, he was more capable of taking care of it than the others."

Similarly, Barbree was told, Grissom would fly second because of his "ability to analyze technical information." Glenn was chosen to fly the orbital mission because "if anyone could bring back a spacecraft, John Glenn could. He had brought back aircraft nobody else could fly, that was shot up so badly in Korea, nobody else could fly."

On May 5, 1961, millions of vicarious astronauts including Benedict, Barbree, and Cronkite, emotionally crammed into *Freedom 7* and joined the pilot on a journey into a world once only dreamed of.

From the Cape, Barbree peered into the sky as *Freedom 7* shrank from impressive rocket, to black and white pencil, to a tiny light against a blue backdrop. The NBC newsman tried to imagine what Shepard was feeling: *He looks so lonely up there, so damn alone.*

In spirit, Shepard was anything but alone. As first the countdown commenced, and then the Redstone spurted off the pad, people everywhere looked to the heavens. They pulled their cars off the highway; they leaped out onto the road, dropped to their knees. Beachgoers crawled away from their Redstone sandcastles and likewise dropped to their knees in prayer. Others clasped and rubbed rosary beads until their hands were sweaty and raw; they lifted the string of spiritual beads to their lips, kissed the shimmering metal cross. Around the nation, churches opened their doors for hopeful parishioners.

When, finally, the 15-minute, 22-second flight had concluded, Walter Cronkite let out a sigh of relief. Then the reporter in him took over: Shepard's flight was just the beginning, he understood, but it was a positive step forward to perhaps, one day, a shot at the Moon. The mission had created a heartwarming sense of romance throughout the nation, particularly at Cape Canaveral. Here was a community of people all of whom were looking toward the stars. Nowhere did you find a down-turned face. Project Mercury wasn't just about beating the Russians, it was about the whole human endeavor.

Benedict had been confident that Shepard's and Grissom's suborbital flights would go pretty much according to plan. The Redstone was a reasonably predictable rocket, and these were, after all, suborbital flights. But when it came time for Glenn to board the often erratic Atlas, Benedict felt a knot tighten in his gut. He had covered previous Atlas launches, and, like Cronkite, had imagined the powerful missile spinning wildly in death throes. But, if successful, Glenn's mission would advance the bridge that was inch by inch reaching toward the Moon.

What Benedict did not envision was the impact the U.S.'s first orbital flight would have on his career.

To Benedict, Barbree and Cronkite, NASA's best sources wore dirty jumpsuits streaked with grease. They were the guys with whom no one talked: the mechanics, the engineers, and the technicians who wielded the wrenches that tightened the screws that converted the sheets of metal, the dozens of connectors and hundreds of feet of wire, into something called a rocket ship. The people who told you why a relay failed, why an engine shut down, why an astronaut needed a window in his capsule. Sometimes, however, the best sources were the least likely of people.

Since arriving at the Cape, Benedict had lived at the Vanguard Hotel, on the ocean front in Cocoa Beach. One day, the desk clerk pulled him aside.

"Howard, we've got a tip that Glenn is coming to church on Sunday," said the clerk. Benedict figured the guy was a pretty reliable source; he was also the deacon at the Presbyterian church Glenn attended.

"Thanks, I'll go down there," replied Benedict.

The reporter felt a chill of anticipation. *Am I the only one with this information? Will I finally be able to land an exclusive?* The timing was perfect to land a big story. The launch had already been postponed seven times, and a cord of tension had wrapped itself around the Cape and the press corps. Benedict crossed his fingers and kept his mouth shut. *Maybe I'll beat out* Life.

Finally, Sunday came. When Benedict arrived at the church, he found it packed to the rafters. Did everyone know Glenn was coming? If they did, it was bad news for Benedict. Other reporters would be on Glenn's trail.

Benedict carried in a folding chair and took a seat in the rear of the church. Then he kept an eye out for the astronaut, all the while praying no newshounds spotted him. Sure enough, minutes later, the astronaut pulled up to the church in his Chevrolet. Somehow, Glenn had surreptitiously sneaked past the hundreds of reporters who spent much of their days racing up and down Route A1A.

Benedict held his breath. He watched Glenn carry a folding chair into the church, then sit down not far from him. Now Benedict had to figure out a way to get to Glenn without scaring off the astronaut.

The reporter's break came at the conclusion of the service, as Glenn stood outside the church, signing autographs. Anxiously, Benedict sidled up to the amiable pilot and began to engage him in small talk. "How's the mission coming along? The Atlas, is it dependable?" Finally, Benedict broached Glenn with the question that would help break the ice between reporter and public hero.

"Psychiatrists are saying that you should be taken off the flight. That you can't do it, you can't stand the pressure."

Glenn looked Benedict straight in the eyes.

"Those are the guys that are crazy, not me," he said flatly.

Benedict smiled. Astronaut and reporter went their own way.

As John Glenn greeted the children emerging from Sunday School, Howard Benedict walked away with his exclusive story, and the quote of a lifetime.

☆ ☆ ☆

Howard Benedict could almost see it happen. Could almost see *Friendship 7* explode in a horrific fireball of flame and twisted metal. *Would today, February 20, 1962, be the day John Glenn finally gets to fly in space?* Benedict wondered. The reporter gazed up at the sleeping Atlas as he had done nearly a dozen times before. *Will Glenn blow up with the rocket?* Benedict could almost see the headline, the tragic story; he could almost see the nation's shock. *Almost . . .*

But Benedict had neither the time nor the inclination to allow his imagination to run amuck. He worked for a wire service that boasted 8,000 members worldwide. So, while other newspaper reporters availed themselves of a late deadline, Howard Benedict was on deadline somewhere across the globe. The reporter went to work formulating a possible story in his mind, one with a happy ending: *John Glenn became the first American to orbit the Earth . . .*

When *Friendship 7* blasted off, Benedict waited until the capsule had reached orbit, then he relayed his pre-conceived article back to the AP's New York bureau. John Glenn *had* proved the psychiatrists wrong: he could in fact bear the pressure of the most daunting U.S. manned space flight to date.

Now NASA was storming forward, headed to Project Gemini, and the astronauts were going to take a part of Howard Benedict with them.

☆ ☆ ☆

Although Benedict had his share of solid background sources, he knew that if he wanted to uncover for his readers the psyches of the astronauts, the hearts and souls of these men, he would have to get to know the astronauts firsthand. That grandiose notion had been made moot by the *Life* contract, but that all changed when the Associated Press decided to install an apartment/office for Howard Benedict at Cocoa Beach's Cape Colony Inn. What was once exclusive information for the magazine would slowly become fair game for Benedict.

Shortly after the flight of *Friendship 7,* Benedict moved out of the old Vanguard Hotel to the splashy Cape Colony.

AP management's edict to Benedict: Do not miss any newsworthy event pertaining to NASA. With that order, Benedict's job became his life. Moving to the Cape Colony provided the breakthrough he needed to cover Project Mercury more intimately. His communication with the crew was no longer restricted to press conferences. Now Benedict was living with the people who were the story.

Slowly, the reporter got to know each of the Mercury pilots; a cursory nod turned into a few minutes of idle chat. Eventually, the astronauts began to venture into Benedict's office, to get the scoop on a breaking story off the news wire. Conversation was pleasant, not strained. Dinners between astronauts and reporter followed. A mutual respect was developing, a friendship growing.

These men were heroes, Benedict acknowledged. But experience had proved to him that the Mercury 7 were not always the all-American boys envisioned by the public. Publicly, they maintained a noble sense of equilibrium, but away from the spotlight they were also little boys at heart. When Jim Rathmann, the owner of a local General Motors dealership, offered sports cars to the astronauts at cut-rate prices, Carpenter, Cooper, Grissom, Shepard, and Slayton selected Corvettes. Schirra took a Maserati. Glenn, consistent to his clean-cut lifestyle, declined, opting to drive his Chevrolet down from his home in Virginia.

Those fast cars were occasionally the source for astronauts' high jinks. On one occasion, Carpenter and Cooper raised Benedict's eyebrows by racing around the parking lot at 3 a.m. Another time, Shepard recounted to Benedict the incident of a Runaway Grissom. The pilot of *Liberty Bell 7* was racing down Route A1A at a speed comparable to that achieved by a Saturn V en route to the Moon. Unfortunately, he and his Corvette were being trailed by a local police officer bent on determining the identity of this high-speed flyer. But this was no contest. The second American in space tramped on the accelerator of the sleek and speedy Corvette and made a beeline back to the hotel, arriving moments before the badly defeated cop. Then the wily Grissom employed his pilot's ingenuity to avoid an undesirable conclusion to his otherwise enjoyable jaunt. Instead of parking his car in a spot outside of his room, he parked in front of

Shepard's. Then Grissom quietly slipped into his room, narrowly avoiding the arriving police officer.

The officer pulled up to the hotel and jumped out of his car. When he saw nobody, he searched for the Corvette that had caused him so much grief. Suddenly there it was, right before his eyes, the very same Corvette he had chased up A1A. He walked up to the car, felt its hood: *hot!* This was the culprit all right.

Determined to bring this ruthless transgressor to justice, the police officer went into the hotel. He learned that the parking spot belonged to Alan Shepard. He found Shepard's room and knocked on the door. He was greeted by a befuddled Shepard, who argued vehemently that he had been in his room during the alleged high-speed race. Eventually, Shepard's persuasiveness prevailed and the policeman went on his way. This was, after all, Alan Shepard, astronaut.

The astronauts did not limit their mischievousness to their sports cars. One time, they hauled a boat into the hotel pool. There they partied until they were finally chased out for making too much noise.

One other night, Cooper got a bucket of fish and tossed them into the pool. He sat there, fishing pole in hand, bottle of beer at his side, and attempted to reel in the fish, one by one. Only one problem: Cooper was unaware that fish aren't particularly hungry after they have swum in chlorinated water. Instead, they prefer to float upside down, on the surface of the water, dead as stones.

This intimate knowledge of the astronauts gave Benedict an edge over most of the other reporters. Manned space flight, he considered, was not about Project Mercury, not about orbits, not about loose heat-shields. It was about the men. They were the story, the people in the headlines. The people about whom Howard Benedict was writing.

☆ ☆ ☆

I *have to keep up with them,* said a voice deep in Jay Barbree's heart. *The others will be in peak physical condition.* Instinctively, he increased his pace. Outer space was beckoning him, after all, and he was competing with 39 other gifted journalists to get there. He had been jogging for 11 years, averaging three to four miles per run, and he considered exercise as much a part of his life as covering space missions, which he had been doing for nearly three decades. But now, at age 53, Barbree was trying to forget about rockets and astronauts and the Moon. He had just returned to his Cocoa Beach home after spending weeks covering the *Challenger* tragedy. The explosion had rocked him personally, as it did every other journalist he knew. Barbree had put aside his personal feelings, however, had rolled up his sleeves, and had gone to work telling America how faulty O-rings in the solid rocket boosters had effected the horrific destruction of the craft and the death of its crew.

But that was behind him now, and Barbree was staring heavenward at the chance of a lifetime: The veteran newsman had a one in 40 chance to do more than just write about space, he had a golden opportunity to write *from* space. But he knew that he had to be in optimum physical condition to have a legitimate shot at being the one person chosen by NASA to be the first journalist to fly in space.

So, as Barbree jogged up the beach, his body starting to tire, his heart banging away in his chest, it was ironic that he did not sense what was about to befall him. He had always assumed that if he overtaxed himself, his body would give him a sign, tell him to slow down, ease up. But there were no extraordinary portents on May 28, 1987. No pain, just extreme fatigue. Barbree blocked out the exhaustion and pressed on.

Barbree's heart was pumping away at 160 beats per minute, racing wildly to provide blood to his brain and his churning legs. He felt his lungs begin to burn, and he grew weaker with each step. His body demanded more blood, more oxygen. His heart worked harder still, but what Barbree did not know was that the coronary arteries that delivered the blood throughout his sagging body were clogged and badly overworked.

Suddenly, without the profound warning he had imagined, his knees began to sag. Then his body went limp. Barbree collapsed to the ground in a heap. He had no pulse, no heartbeat. Barbree was clinically dead, his world a black dreamless sleep.

☆ ☆ ☆

Benedict, Cronkite, Barbree: They were the stick and rudder test pilots of the journalism world, who used words and images to fly Americans into low Earth orbit, or to the craggy landscape of our nearest celestial neighbor.

Like their Mercury 7 counterparts, each reporter considered himself the most deserving of the title "First Journalist In Space," the Al Shepard of the Fourth Estate. The only problem was, while NASA's original pool of eligible test pilots totaled approximately 500, the number of journalists who applied for NASA's Journalist in Space Program was about 1,700. Imagine 1,700 Deke Slaytons, John Glenns, and Scott Carpenters vying for one and only one Mercury mission. Barbree tried to ignore such overwhelming odds, as did Benedict, Cronkite, Pelley, and the scores of other news reporters eager to bring a microphone, camera, tape recorder, or notebook into a weightless newsroom.

I would be credible for them, thought Barbree, who had worked for NBC for some 30 years, and had written numerous articles and a few books on space exploration. *I can do a radio report and write a newspaper story. I can report back to the readers and listeners on Earth what it was like to be in space. No engineer can do that. If anybody has paid his dues, I have.*

Barbree was right. By the time NASA announced in the fall of 1985 that a journalist would fly a Space Shuttle mission as part of its Space Flight Participant Program, Barbree had covered every aspect of manned space flight, every launch of every manned Mercury, Gemini, Apollo, Skylab, and Shuttle mission. He had proved to his profession and to the world that he could effectively and succinctly articulate the technological and emotional aspects of flight. Based on those criteria he was an ideal candidate for the program. According to NASA, the journalist chosen should be able to observe and record the mission's events, as well as his or her feelings about the flight. In a sense, the person chosen would be proxy for every fan of space flight in the world.

But, like Alan Shepard in 1959, Barbree had rivals who were just as quick with their deadline reports, just as eloquent with their prose, just as meticulous. Few possessed those journalistic attributes as admirably as Benedict. All the hours spent with Wernher von Braun, drinking German beer at his hotel and discussing space exploration, had paid off; Benedict knew the hardware of space flight, and he knew the voyagers.

And Cronkite? When it came to national television coverage, there was Cronkite and there was everyone else. Cronkite gave you the story, devoid of mawkishness. He was able to separate his emotions from the story. That, in and of itself, enhanced his unmistakable credibility.

All three newsmen yearned for the assignment of a lifetime. All three were convinced they would be safe aboard a Shuttle. All had faith in NASA. But it was a long shot for each, just as it was a long shot for Al Shepard to garner the coveted ride in *Freedom 7.*

The Journalist in Space program was part of NASA's Space Flight Participant Program. More than a year earlier, classroom teacher Christa McAuliffe had been the first candidate selected from a field of more than 11,000 teachers. She was scheduled for a January, 1986 mission.

At the request of NASA, the Association of Schools of Journalism and Mass Communication (ASJMC), would coordinate the selection process. The enormous group of candidates would be pared down to 100, based on the strength of essays submitted with the applications. Panels of journalists and professors from universities would review applications, essays, letters of recommendation, and would hold interviews to trim the total to 40 national semifinalists. That group would travel to Washington, D.C., to be interviewed by a 14-member selection panel that would select the five finalists. Those five journalists would then go to Houston's Johnson Space Center to undergo physical examinations and receive briefings on space flight. Finally, a seven-person committee would judge the finalists based on compatibility with the flight crew and an aptitude for the flight training. The selected candidate and his or her backup would then undergo extensive training, similar to that experienced by the NASA astronauts.

For Barbree, Cronkite, and Benedict, the long climb to outer space began with a common employment application which requested biographical information, employment history, and a professional recommendation. Barbree read the rules of the application: "Starting with your present position, describe your professional experience by completing the sections below. Identify and describe a total of at least five years of journalistic experience and any related experience you may wish to have considered."

He panned down the page and saw a box, some two inches high by seven inches wide. In there, he was to recount his basic responsibilities and accomplishments of a career that spanned more than a generation.

"That's bullshit," he snorted, then chucked the application into the garbage can. *They give you no freedom at all, they put you in a straight-jacket.* Eventually, Barbree cooled down, filled out the application, squeezed every last word into that compact box, and sent it in.

Things did not work out favorably for Benedict. Like Barbree, he had spent many grueling weeks reporting on the *Challenger* explosion and subsequent investigation. And like Barbree, he was mentally and physically tired from the extensive work he had done on the story. Unlike his colleague, when it came to reviewing the application, he was less than meticulous. Applicants were required to submit with their application two samples of their professional work. Benedict, however, failed to read the rules correctly. He submitted three copies of a book he had authored, along with three copies of a feature story, and one news story he had written.

The oversight was costly. By failing to follow the rules, Benedict was automatically disqualified from the program. Conversely, when the 100 regional national semifinalists were named, Barbree, Cronkite, and Scott Pelley had all made the cut.

As a precursor to the spring, 1986, interview process that would narrow the pool to 40 candidates, NASA and ASJMC asked the remaining candidates: "In light of the Challenger disaster, are you sure you wish to do this?"

☆ ☆ ☆

Christopher Glenn, who covered *Challenger* and many other missions for WCBS radio in New York City, had once pondered the possibility of flying in space. But with each Shuttle mission he covered, he grew increasingly concerned about the structural integrity of the spacecraft and its solid rocket boosters. When *Challenger* exploded right before his eyes, Chris Glenn decided that NASA needed to find a newer and safer way to launch a man into space.

The Shuttle explosion would haunt the respected journalist for years to come. Although his enthusiasm for space flight would not diminish, with STS-26, the 1988 follow-up to *Challenger*, and each subsequent mission, Glenn was seized by fear with every liftoff. The unbridled elation he had felt at the moment of ignition and liftoff was supplanted by a trembling, gut-wrenching specter of death: *The damn thing is gonna blow up in front of my eyes again.*

Unlike Glenn, whose skepticism of the hardware quashed his interest in the Journalist in Space Program, Pelley, Barbree, and Cronkite each gave NASA the thumbs-up. Each would express his unabashed interest in flying in space during interviews with regional selection panels.

Barbree had often pondered the role a journalist could play in space exploration. He had even approached NASA with some plans and ideas for a mission that would carry a reporter. "See if you can find me a free radio channel that won't interfere with the telemetry and voice channel coming back from the crew to Mission Control."

Barbree postulated that with this open transmission loop, he could provide an ongoing narration of his feelings, sensations, and thoughts of the flight, from ignition to reentry.

"If I'm sitting in the seat, I want to be able to say 'Ohhh, boyyyy, here we goooo!' to describe the sensations of flight," he said.

If liftoff was a rocky ride, shaking him through his bones and into his soul, Barbree wanted to tell the world *I'm scared shitless.*

He advanced his case before the regional selection panel. During the interview process, Barbree proposed another project for the mission.

"I'd like to do a book for children, a book for young adults, and a book for adults on the flight."

One skeptical, highbrow journalist/professor shot Barbree a doubtful look.

"How the heck are you going to do a book for small children and make them understand space?" the professor asked.

"Well, I'll take up Slinky toys and other toys that they were used to playing with, and I would show them how they operate in space. Then they can understand what a yo-yo does in space, but it does something different here on gravity. I could do it pictorially."

The professor squinted thoughtfully. "Oh, yeah, that's a good idea," he admitted.

Apparently, the rest of the selection committee members agreed. Barbree was chosen as one of 40 national semifinalists. Cronkite was also selected, along with other respected news reporters such as Morton Dean and Lynn Sherr. For Benedict and Pelley, the journey ended just a few miles short of outer space.

Little did the 40 remaining journalists know that their dreams of reporting from a spacecraft were about to be shattered as well. In the aftermath of the January 28, 1986, Shuttle explosion, NASA administrators had begun questioning the prudence of sending another civilian into space, particularly since it likely would be some time until the resumption of flights. That same month, the agency had put the Journalist in Space Program on hold, only to resume it some two weeks later with a revised selection schedule and process. But, on July 1, 1986, two weeks after the Journalism Advisory Committee had recommended the completion of the selection process, NASA once again advised the Association of Schools of Journalism and Mass Communications to stop the program.

The project was not officially canceled. At least that's what NASA administrators intimated to Barbree. Perhaps in a year or so, a definite date could be chosen for a journalist to fly in space.

☆ ☆ ☆

When Barbree began his fateful workout nearly a year later, he still believed a newsperson would one day fly in space. He ran with that goal clear in his mind. And with a home on Cocoa Beach, Barbree could imagine no more soothing environment in which to train, to forget the horrific events of the last few months: the explosion of the *Challenger*, the subsequent recovery of the crew's bodies. Now it was time to be peaceful.

But as Barbree's mind began to relax, his body fought vainly against a disease that had been waiting to seize him. He raced past the Baptist church and headed for a finish line he had established in his mind. He reached it 27 minutes after he had started his run. Not bad, he told himself.

Barbree turned and began to walk back to his home. He was lathered in sweat. He was utterly exhausted. As he walked, he looked up at his house, just in the distance. He felt no pain. But suddenly, Barbree's heart began to spasm wildly. Barbree knew something was wrong.

First his legs went, then the rest of his body followed.

To the many beachgoers, Jay Barbree appeared to be just another sunbather, relaxing in the warm Cocoa Beach sand. But RCA engineer David Frank suspected otherwise. Earlier, Frank had taken his daily walk, and had been passed by Barbree. Now, as he headed toward the motionless form on the beach, Frank sensed something was terribly wrong. He raced over to Barbree, felt for a pulse. Nothing. Frank, who was certified in CPR, knew he had just four to six minutes to get oxygen to Barbree's brain. Otherwise, there was little chance of reviving the stricken man. He worked feverishly to resuscitate Barbree. He shouted for help.

The commotion roused the interest of passersby. It also caught the attention of Pat Sullivan, a college student who worked at a nearby restaurant. Sullivan, who was also trained in CPR, sprinted out to the beach to offer his assistance. He was stunned when he recognized the unconscious man was Jay Barbree. Sullivan had been friends with Barbree's daughter, Karla, since high school.

After the initial shock had passed, Sullivan joined David Frank in what was slowly becoming a futile effort to save Jay Barbree's life.

☆ ☆ ☆

At the time of Barbree's collapse, Debi Hall was cooking dinner for her husband in their third-story home in Park Place Condominium. Hall, an emergency medical technician, served as the EMT for the astronauts who lived or trained in the area. She was also married to a member of the Cocoa Beach police department. As fate would have it, the buzz of her husband's police radio had caught her attention. When she heard that a man had collapsed on the beach, Hall walked out to her balcony. She spotted a small crowd gathered around a person who lay motionless in the sand.

Hall raced out of her home and hurried down to the spot where Barbree lay. His eyes were open, empty. His face was purple, lifeless. She took over from Frank and Sullivan.

As Hall applied CPR to Barbree, a rescue squad arrived, carrying Lee Proctor and paramedic Ed Clemons. The two men jumped out of the wagon and joined Hall. Proctor inserted an airway tube into Barbree's throat, then he and Clemons attached an oxygen bottle to the tube. They began to fill Barbree's lungs with oxygen. Proctor continued the chest compressions he had begun moments earlier. Slowly, the team began to manually move blood through Barbree's body. Just as slowly, Barbree's color began to return, and a few times he attempted to breathe on his own.

Moments later, an ambulance arrived. EMT Chris Bedard leaped out of the vehicle, carrying a defibrillator pack. He checked Barbree's pulse. Nothing.

Bedard advised the others to step back; then he applied the defibrillator paddles to Barbree's chest. The electrical shock made Barbree's body leap a few inches off the ground. Still no pulse or heartbeat. Bedard tried again. No response. He tried one more time, sending 360 joules of shock into Barbree's lifeless body.

Nothing.

Undaunted, Bedard and the other EMTs loaded Barbree into the ambulance, and the team set off for Cape Canaveral Hospital.

Nearly 20 minutes had passed since Barbree's attack, but Bedard refused to give up. "I'm going to hit him one more time," Bedard announced. Then he sent another wave of electricity through Barbree's body.

Suddenly, a heartbeat. Not regular. But it was a start. On the sixth attempt, Barbree began breathing on his own.

☆ ☆ ☆

Space flight inspires confidence, it nourishes the soul, it kills.

It does not provide the answers to life and death, however, as Barbree concluded. Before "dropping dead," Barbree had believed in God in a nebulous, non-denominational way. Twenty minutes of non-existence failed to confirm an afterlife or the true nature of God: Barbree experienced no out-of-body flight, no rendezvous with a heavenly angel, no joyous procession into a tunnel of light. There was no grand epiphany.

The spiritual changes in Barbree were more subtle. After the heart attack and subsequent surgery, Barbree intuitively believed that there was something beyond his earthly life, an existence that transformed sleep without dreams into a nirvana.

I believe there is a supreme force . . . I simply do not know what's out there, he thought.

What's out there, others have told him, may soar a man beyond the Moon, beyond the stars. Beyond one's human-bound senses. Take Jim Irwin, lunar module pilot of Apollo XV, NASA's most ambitious flight to date. During the crew's 67-hour stay on the lunar surface, astronauts Irwin and Dave Scott unveiled the first powered vehicle driven on the Moon—the Lunar Roving Vehicle. While on lunar expeditions, the astronauts rode the Moon car to the cusp of Hadley Rille, a 1,200-foot-deep canyon. There, they collected rock samples, including a piece of anorthosite that was over four billion years old, older than any rocks ever found on Earth. After the flight, the rock was dubbed The Genesis Rock.

Weeks after Apollo XV had been relegated to the nation's collective memory, Barbree and Irwin discussed the flight. As the friends talked, Irwin disclosed to Barbree that while on the Moon he had been instructed by God to seek out the Genesis Rock. Upon His command, the astronaut redirected the lunar vehicle from its predetermined course. When the Lunar Rover finally stopped, the Genesis Rock sat before Irwin.

A few months after Irwin's death in 1991, Barbree was sitting in his living room watching a Saturday afternoon college football game. The last person on his mind was Jim Irwin.

Suddenly, Jay Barbree felt his jaw tighten. It felt as if it would burst. He was overwhelmed by an overpowering conviction that his buddy,

Irwin, was inside of him. *My body's going to burst right out and Jim's going to fly right out of me!* he thought.

"Brother Jim, what is it?" Barbree called.

Call Mary, tell her I'm okay, Barbree heard Irwin tell him.

"Brother Jim, I'll call Mary," Barbree agreed, as if Irwin were with him in the room.

Just as inexplicably, Barbree's body relaxed, his jaw slackened. He turned off the television and went to the kitchen where Jo, his wife, gazed quizzically at him.

"Jim Irwin just came to me," Barbree exclaimed. "He wants me to call Mary and say everything's all right."

Jo Barbree looked at her husband as if he had just returned from Saturn.

"I'm not kidding, Jo, he was here."

Barbree ignored his wife's skepticism and went to call Mary Irwin, Jim's widow. When he received her answering machine, he left a message. It would take a grieving Mary Irwin a few months to return his call.

Brother Jim is all right, Barbree knew. *He's all right.*

☆ ☆ ☆

It took Jay Barbree a few years to abandon his dream of flying in space. He agrees with Cronkite, who called NASA's decision to scrub the program a "knee-jerk reaction to the *Challenger* disaster." Now 64, Barbree will continue to tell stories about space flight. Two months before John Glenn's Shuttle flight, the veteran reporter sensed that a part of him would fly with the longtime senator from Ohio.

Forty years of covering space flight have opened Barbree's mind and soul to the possibility of a life beyond the huge, fragile satellite called Planet Earth. Barbree is, by profession and nature, an inquisitive person. By asking countless questions, by extricating the truth from fallacy, Barbree has broken story after story. Outer space has been a long, dark corridor connecting this physical world to another existence just beyond experience: Jim Irwin's Genesis Rock world;his own black, sleepless world.

A dimension far beyond Earth.

Do we go from this dimension to another dimension when we leave this body? Barbree wonders. *Is this body nothing more than a vehicle for us on Earth?*

Jay Barbree shrugs his shoulders.

I do not have the answers, he concludes.

Heroes, 40 Years Later

I was like a 4-year-old clutching my *Freedom 7* turkey baster the day I heard Alan Shepard had died. On July 22, 1998, less than 24 hours after America had lost Shepard, I slipped back in time, from age 39, and reconnected with every moment I had yearned to become an astronaut. Maybe we all did.

During the course of writing this book, I had frequently wondered how I would feel if one of the remaining Mercury astronauts passed away. Now I knew. When I heard the news, my kids—Brooke, now six weeks shy of her fourth birthday, Jordan, now 6, and Adam, nine months—were seated in a semicircle around the television set, watching one of the countless children's shows they enjoyed. It was dinner hour, and, typically, at this time we would switch to the local news.

When I flipped from the PBS station to New Haven's Channel 8, a picture of Alan Shepard, circa the Mercury years, flashed across my screen. It blinked off a second later, but not before I felt my stomach sink. A wistful-looking weatherman quickly replaced the snapshot of Shepard.

"I was 6 when Alan Shepard flew in space," he said. "I remember it vividly."

Those were the last words I heard him utter.

Debbie, my wife, turned and looked at me. "Do you think he died?"

"I have an awful feeling," I responded, not wanting to know the truth.

I trudged warily to the dining room, then booted up my computer. Moments later I was on the Internet and scanning the day's news stories.

"Alan Shepard dies," read the first news brief.

"Oh, no!" I moaned.

"Is he dead?" asked Debbie.

"Yup, he had been battling leukemia. Somehow, he was the last one I expected to go next."

I shut off the computer and walked out to the deck, where my family and I had once mused about the stars and space flight. I felt an emotional void spreading inside of me, as if I had just lost a loved one. In a way, the death of Alan Shepard was equally profound: *I thought these guys would live forever*. Deep down inside of me, that 4-year-old little boy who had grown up with space flight really believed the Mercury guys would live forever.

If Alan Shepard is mortal, then so am I.

Five-and-a-half months shy of my 40th birthday, I suddenly felt 80. And yet, that innocent little boy in me was still reaching out, trying desperately to connect with a superhero rendered mortal by aberrant blood cells that just didn't understand that astronauts can't die, they just fly in space forever. I stood, frozen, by the sliding glass doors and thought: *I wonder if he's flying around right now. I wonder if he might be floating over us on his way back to the Moon.*

Shaking my head, I tried to purge the numbness I felt from his loss. It was no use. I gazed into the kitchen, where I had stood the night Shepard called me. Where I had held a dripping eight-ounce saucepan when the shrill ring of the phone launched me to the ceiling. Where Shepard's relaxed voice had made me feel as if I had known him forever.

You don't expect the first American to fly in space to call you while you are washing pots and pans, I thought now. I walked over to the small kitchen window and gazed wistfully into my backyard. I had been doing the same thing the night Shepard and I spoke. Back then, however, my mouth was full of sawdust. This was Alan Shepard I was talking with!

Now he was gone, and my mouth was again all dry, this time from sadness. His legacy to me was a genial telephone interview that lasted a lifetime. I turned from the window and headed to my bedroom, where I stored all the recordings of my interviews. I wanted to hear Al's voice just one more time.

In the days following Shepard's death, people asked me what I thought of the heroic astronaut.

"Al Shepard was every person who turned a wild dream into reality," I replied.

What amazed me most about Shepard and nearly all the other astronauts was how each held a unique perspective on space exploration. Did Al Shepard, Bob Crippen, Vance Brand, Steve Smith, Gordon Cooper, and the dozens of other astronauts share my romantic perspective on space flight?

Vance Brand flew with Deke Slayton on the 1975 Apollo-Soyuz Test Project. He was also a Capcom on the Apollo XIII flight. Today, Brand is assistant director of flight operations at NASA's Dryden Flight Research Center. There are moments, however, when he thinks back to the historic international mission when representatives of erstwhile enemy nations dined together in capsules interlocked in outer space. That's when Brand returns to the last night of the odyssey, when he shared a final dinner with crewmates Tom Stafford and Slayton. *How I hate to come home,* he thought, and then expressed that sentiment to his friends. *Maybe I wouldn't want to stay up here forever, but I'm getting so much out of this flight. Maybe, just a little bit longer?*

Brand relishes the memory now, but that wistful feeling dulls quickly. He is, after all, at his desk at this moment, working in flight research on the aircraft of the future. Brand searches for the mental switch that he activates from time to time, the emotional toggle that redirects his mind, intellect, and emotions away from the romance of flight, back to the job at hand. Back to a grounded, earthly reality.

Forget how much you'd like to be flying, he tells himself, *and just think about your new responsibilities and objectives.*

Then Brand flips the switch in his brain.

Apollo-Soyuz vanishes.

They successfully flew in space, these dozens of astronauts, because they knew when to flip that psychological switch. When to be emotional observer, when to be emotionless pilot. Shepard captured the polarized responses to space flight, more than 35 years after his *Freedom 7* flight.

"We had six flights, we had some things go wrong, but from a piloting point of view, everything went extremely well. It showed pilots can, under certain circumstances, react very favorably, and I think that was something a lot of people criticized," said Shepard a year before his death.

Did their early missions send a message to the world? Shepard was not certain.

"We were engineers, we were test pilots, and we basically looked at the whole program from the standpoint of improving technology in all areas, not only in rockets or electronics, but propulsion systems and everything, rather than bringing some divine message back from lunar orbit."

An overly humble and mechanical appreciation of the space program? Perhaps, on the surface. But underscoring Shepard's appreciation of the technology was a deep respect for the personnel involved and for the accomplishments of his *Freedom 7* and Apollo XIV missions. Not only did he lift America's sinking morale off the ground with his suborbital flight in 1961, and his 1971 Moon landing, but Shepard also helped restore the nation's trust in NASA just months after the nearly disastrous Apollo XIII.

In reflecting on the historic significance of his flights, the emotional observer in Shepard surfaced, if for just a few precious moments.

"It gave me a sense of satisfaction, and I don't mean gloating about it in any sense. It was a personal feeling, if whatever I wanted to do in the world I could do, if I just set my mind to it."

If, publicly, Shepard yielded grudgingly to his emotions, Gordon Cooper waxes poetic over his flights aboard *Faith 7* and Gemini V. The first astronaut to manually fly his spacecraft through reentry, today the 71-year-old heads Galaxy Aerospace, a company that does research and development work on aircraft, through to production development. Always mechanically inclined, Cooper eased through the often difficult transition from public life to private life.

Nevertheless, ask him about Project Mercury and he becomes a kindly Oklahoma grandfather recounting humorous and touching anecdotes to his grandkids.

Take, for instance, the occasion Cooper was questioned by the Internal Revenue Service for listing the wrong address on his income tax form.

"You listed 350 days a year of travel," an agency representative noted.

"You find me one address that I have the most time at, rather than the address I picked," responded Cooper.

"You're right."

Cooper and the other Mercury guys lived out of suitcases and worked 28 hours a day, nine days a week on a program in which everyone was a rookie. What made it all worthwhile? Twenty-two orbits, 22 breathtaking sunrises and sunsets during the flight of *Faith 7*. A day to ponder the beauty and destruction of a world suddenly so distant.

"I think we came from our flight realizing what a big, beautiful world this is," said Cooper. "You don't see any boundaries up there, and you begin to wonder why we have all these boundaries and all these battles between people, and why we can't learn to get along a little better and have a bit more peaceful existence."

☆ ☆ ☆

From outer space Earth is a lonely place, Wally Schirra concluded, during his flights aboard *Sigma 7,* Gemini VI, and Apollo VII. *We have to take care of this spaceship; there's no place else to go.*

During his Mercury mission, Schirra peered back at that solitary world, marveled at the fragile egg-shell atmosphere that contained so many colors. The astronaut was moved to record the captivating image; unfortunately, however, he did not have a color camera on board. After considering the problem for a few moments, Schirra did what any 4-year-old kid would have done. He broke out paper and crayons and rendered his own sketched rendition of his planet's atmosphere. After the flight, that drawing proved to be of great interest to astronomers.

Being alone, more than a hundred miles from home, made Schirra feel so small, so insignificant, compared to the immenseness of space. Again and again, Schirra watched his home planet awaken and then fall back to sleep. He noted the wormlike canals across the landscape, the tiny specks of lights. Entire cities asleep, then at work, then at play. He was left with a sobering sense of mortality, versus the immortality of God.

As the years passed, Schirra concluded: *Science couldn't handle all these things alone; it had to have some deity over it.* And over those years, as first Gus, then Deke, and finally Al soared through Earth's gravity, beyond Earth orbit, and into the hands of God, Wally Schirra realized that, like the space voyages of the Mercury 7, all things come to an end.

Bibliography

Dewaard, E. John and Nancy. *The History of NASA: America's Voyage to the Stars.* New York: Exeter Books. 1984.

Allen, Joseph P. with Russell Martin. *Entering Space: An Astronaut's Odyssey.* New York: Stewart, Tabori, and Chang. 1984.

Yenne, Bill. *The Astronauts: The First 25 Years of Manned Space Flight.* New York: Exeter Books. 1986.

Shepard, Alan and Deke Slayton with Jay Barbree and Howard Benedict. *Moon Shot: The Inside Story of America's Race to the Moon.* Atlanta, GA: Turner Publishing. 1994.

"We Seven." The Astronauts Themselves (M. Scott Carpenter, L. Gordon Cooper, Jr., John H. Glenn, Jr., Virgil I. Grissom, Walter M. Schirra, Jr., Alan B. Shepard, Jr., Donald K. Slayton). New York: Simon and Schuster. 1962.

Chaikin, Andrew. *A Man on the Moon: The Voyages of the Apollo Astronauts.* New York: Penguin Books. 1994.

Neal, Valerie, and Cathleen S. Lewis, and Frank H. Winter: *SPACEFLIGHT - The Complete Illustrated Story - from Earliest Designs to Plans for the 21st Century.* New York: Macmillan USA. 1995.

Aldrin, Buzz and Malcolm McConnell: *Men From Earth.* New York: Bantam Books. 1989.

Life In Space. Alexandria, VA. Time Life Books. 1983